A Course in
Field Theory

A COURSE IN
FIELD THEORY

Pierre van Baal

CRC Press
Taylor & Francis Group
Boca Raton   London   New York

CRC Press is an imprint of the
Taylor & Francis Group, an **informa** business

CRC Press
Taylor & Francis Group
6000 Broken Sound Parkway NW, Suite 300
Boca Raton, FL 33487-2742

© 2014 by Taylor & Francis Group, LLC
CRC Press is an imprint of Taylor & Francis Group, an Informa business

No claim to original U.S. Government works

Printed on acid-free paper
Version Date: 20130627

International Standard Book Number-13: 978-1-4665-9459-3 (Hardback)

DOI: 10.1201/b15364

Visit the Taylor & Francis Web site at
http://www.taylorandfrancis.com

and the CRC Press Web site at
http://www.crcpress.com

Contents

Introduction

Field theory is most successful in describing the process of scattering of particles in the context of the standard model, and in particular in the electromagnetic and weak interactions. The Large Electron Positron (LEP) collider operated from 1989 until 2000. In a ring of 27 km in diameter, electrons and positrons were accelerated in opposite directions to energies of approximately 45 GeV. This energy is equivalent to half the mass (expressed as energy through $E = mc^2$) of the neutral Z^o vector boson mass, which mediates part of the weak interactions. The Z^o particle can thus be created in electron–positron annihilation at the regions where the electron and positron beams intersect. As a Z^o can be formed out of an electron and its antiparticle, the positron, it can also decay into these particles. Likewise it can decay in a muon–antimuon pair and other combinations (like hadrons). The cross section for the formation of Z^o particles shows a resonance peak around the energy where the Z^o particle can be formed. The width of this peak is a measure of the probability of the decay of this particle. By the time you have worked yourself through this course, you should be able to understand how to calculate this cross section, which in a good approximation is given by

$$\sigma = \frac{4\pi \alpha_e^2 E^2/27}{(E^2 - M_{Z^o}^2)^2 + M_{Z^o}^2 \Gamma_{Z^o}^2}, \tag{1}$$

expressed in units where $\hbar = c = 1$, $\alpha_e = \frac{e^2}{4\pi} \sim 1/137.037$ is the fine-structure constant, E is twice the beam energy, M_{Z^o} the mass and Γ_{Z^o} the decay rate (or width) of the Z^o vector boson. The latter gets a contribution from all particles in which the Z^o can decay, in particular from the decay in a neutrino and antineutrino of the three known types (electron, muon, and tau neutrinos). Any other unknown neutrino type (assuming their mass to be smaller than half the Z^o mass) would contribute likewise. Neutrinos are very hard to detect directly, as they have no charge and only interact through the weak interactions (and gravity) with other matter. With the data obtained from the LEP collider (Figure 1 is from the ALEPH collaboration), one has been able to establish that there are *no* unknown types of light neutrinos, i.e., $N_\nu = 3$, which has important consequences (also for cosmology).

The main aim of this field theory course is to give the student a working knowledge *and understanding* of the theory of particles and fields, with a description of the standard model towards the end. We feel that an essential ingredient of any field theory course has to be to teach the student how Feynman rules are derived from first principles. With the path integral

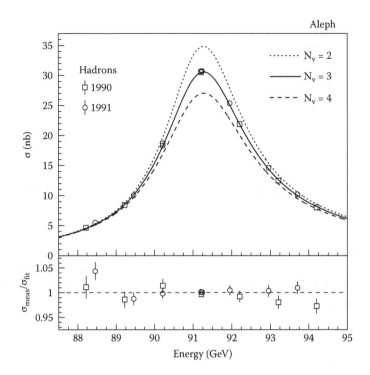

FIGURE 1

Comparison of standard model predictions to the observed cross section $e^+e^+ \to$ hadrons at the Z resonance. The lower plot shows the ratio of the measured cross sections and the fit. Credit: CERN.

approach this is feasible. Nevertheless, it is equally essential that the student learns how to use these rules. This is why the problems form an integral part of this course. As Julius Wess put it during his course as a Lorentz professor at our institute, 'You won't become a good pianist by listening to good concerts.'

These lecture notes reflect the field theory courses I taught in the fall of 1992 at Utrecht, and in 1993, 1994, 1996, 1998, and 2000 at Leiden. I owe much to my teachers in this field, Martinus Veltman and Gerard 't Hooft. As I taught in Utrecht from 't Hooft's lecture notes "Inleiding in de gequantiseerde veldentheorie" (Utrecht, 1990), it is inevitable that there is some overlap. In Leiden I spent roughly 25 percent longer in front of the classroom (three lectures of 45 minutes each for 14 weeks), which allowed me to spend more time and detail on certain aspects. The set of problems, 40 in total, were initially compiled by Karel-Jan Schoutens with some additions by myself. In their present form, they were edited by Jeroen Snippe.

Of the many books on field theory that exist by now, I recommend the student to consider using *Quantum Field Theory* by C. Itzykson and J.-B. Zuber (McGraw-Hill, New York, 1980) in addition to these lecture notes, because it offers material substantially beyond the content of these notes. I will follow to a large extent their conventions. I also recommend *Diagrammatica: The Path to Feynman Diagrams,* by M. Veltman (Cambridge University Press, 1994), for its unique style. The discussion on unitarity is very informative, and it has an appendix comparing different conventions. For more emphasis on the phenomenological aspects of field theory, which are as important as the theoretical aspects (a point Veltman often emphasised forcefully), I can recommend *Field Theory in Particle Physics* by B. de Wit and J. Smith (North-Holland, Amsterdam, 1986). For path integrals, which form a crucial ingredient of these lectures, the book *Quantum Mechanics and Path Integrals* by R.P. Feynman and A.R. Hibbs (McGraw-Hill, New York, 1978) is a must. Finally, for an introduction to the standard model, useful towards the end of this course, the book *Gauge Theories of Weak Interactions,* by J.C. Taylor (Cambridge University Press, 1976), is very valuable.

And More

Gerard 't Hooft finally wrote a summary of his lecture notes (192 in www. phys.uu.nl/~thooft/gthpub.html, December 23, 2004). It is so good that I must quote it here: Gerard 't Hooft, 'The Conceptional Basis of Quantum Field Theory,' in *Handbook of the Philosophy of Science, Philosophy of Physics,* eds. J. Butterfield and J. Earman (Elsevier, Amsterdam, and Oxford, 2007), Part A, pp. 661–729.

Of course I continued to give lectures on field theory, and taught it also in 2002, 2004 and 2007. But I had a stroke on July 31, 2005. I recovered to such an extent I could lecture again for two years; unfortunately some new complications prevent me from teaching at present. This 'And More' is written in December 2012, but the remainder of this course (including numerous corrections) was written before July 2005. Only one thing was corrected during the 2007 course: $\tilde{\pi}(\vec{k})$ was interchanged with $\tilde{\pi}^*(\vec{k})$ in the equation that defines $a(\vec{k})$ and $a^\dagger(\vec{k})$ in Equation (2.7).

Recently the masses of neutrinos have been more accurately determined, but I have not updated that (because it would need more discussion). And finally, the LEP collider at CERN was replaced by LHC (Large Hadron Collider), which circulates protons in either direction. They have found (July 4, 2012) a particle that seems to be the Higgs at roughly 126 GeV. If true, this completes the standard model, but that there is something beyond it is already clear.

Acknowledgments

These lecture notes were available in pdf, and I did not bother much to turn them into a book. But at the end of July 2005 I had a stroke. Nevertheless, I did teach again (in a modified format) and I want to thank Jasper Lukkezen,

Louk Rademaker, Jorrit Rijnbeek and Jörn Venderbos for taking part, and Aron Beekman for helping and rating the problems. I could not continue with teaching, but I never give up.

Then, on November 13, 2012, Dr. John Navas, a physics senior acquisitions editor at Taylor & Francis/CRC Press, approached me after he came across my lecture notes. Would I publish this? The books they have are good, and I worried a bit about how long I could maintain www.lorentz.leidenuniv.nl/~vanbaal/FTcourse.html. They could process the LaTeX file, but I made sure that he knew I had a stroke. But he continued with the publication, and I am extremely grateful for it! The last message sent on January 15, 2013, ends with: "It was extremely exciting to work with you, and you have already done much of the work to turn it into a book." This was his last day and he was moving on to something else, so he might not have seen it. That is why I say it again. Thank you John!

This job was taken over by Francesca McGowan, and I also thank her. I am also grateful to Marcus Fontaine for coordinating the production and being so flexible. I doubted, but I managed to correct the proofs at the deadline, and before you proudly lies *A Course in Field Theory*.

1

Motivation

DOI: 10.1201/b15364-1

Field theory is the ultimate consequence of the attempts to reconcile the principles of relativistic invariance with those of quantum mechanics. It is not too difficult, with a lot of hindsight, to understand why a field needs to be introduced. This is not an attempt to do justice to history—and perhaps one should spare the student the long struggle to arrive at a consistent formulation, which most likely has not completely crystalised yet either—but the traditional approach of introducing the concept is not very inspiring and most often lacks physical motivation. In the following discussion I was inspired by *Relativistic Quantum Theory* from V.B. Berestetskii, E.M. Lifshitz, and L.P. Pitaevskii (Pergamon Press, Oxford, 1971). The argument goes back to L.D. Landau and R.E. Peierls (1930).

An important consequence of relativistic invariance is that no information should propagate at a speed greater than that of light. Information can only propagate inside the future light cone. Consider the Schrödinger equation

$$i\hbar\frac{\partial\Psi(\vec{x},t)}{\partial t} = H\Psi(\vec{x},t). \tag{1.1}$$

Relativistic invariance should require that $\Psi(\vec{x},t) = 0$ for all $(\vec{x},t)$ outside the light cone of the support $N_\Psi = \{\vec{x}|\Psi(\vec{x},0) \neq 0\}$ of the wave function at $t = 0$, Figure 1.1.

Naturally, a first requirement should be that the Schrödinger equation itself is relativistically invariant. For ordinary quantum mechanics, formulated in terms of a potential

$$H = \frac{\vec{p}^2}{2m} + V(\vec{x}), \tag{1.2}$$

this is clearly not the case. Using the relation $E^2 = \vec{p}^2c^2 + m^2c^4$, the most obvious attempt for a relativistically invariant wave equation would be the Klein–Gordon equation

$$-\hbar^2\frac{\partial^2\Psi(\vec{x},t)}{\partial t^2} = -\hbar^2c^2\frac{\partial^2\Psi(\vec{x},t)}{\partial\vec{x}^2} + m^2c^4\Psi(\vec{x},t). \tag{1.3}$$

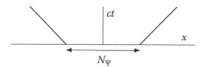

FIGURE 1.1
The (future) light cone for N_Ψ.

However, for this equation the usual definition of probability density is not conserved

$$\partial_t \int d_3 \vec{x} \Psi^*(\vec{x}, t) \Psi(\vec{x}, t) \neq 0. \tag{1.4}$$

As this is a consequence of the fact that the equation is second order in time, this can be easily remedied, it seems, by taking the 'square root' of the Klein–Gordon equation

$$i\hbar \frac{\partial \Psi(\vec{x}, t)}{\partial t} = \sqrt{\left(-\hbar^2 c^2 \frac{\partial^2}{\partial \vec{x}^2} + m^2 c^4 \right)} \Psi(\vec{x}, t). \tag{1.5}$$

We shall show that this, however, violates the principle of causality, i.e., the wave function propagates outside of its light cone, which is unacceptable. Nevertheless, we will learn something important from that computation, namely that negative energies seem unavoidable when trying to localise wave functions within the light cone of N_Ψ. But first we will provide a simple heuristic argument based on the uncertainty relation.

From the uncertainty principle $\Delta x \Delta p > \hbar/2$ and the bound on the speed involved in any measurement of the position, it follows that precision of a measurement of the momentum is limited by the available time $\Delta t \Delta p > \hbar/c$. Only for a free particle, where momentum is conserved, would such a measurement be possible, but in that case, of course, the position is completely undetermined, consistent with the plane wave description of such a free particle (the light cone of N_Ψ would in that case indeed give us no constraint). More instructive is to look at how accurately we can determine the position of a particle. As the momentum is bounded by the (*positive*) energy ($p \leq E/c$) and as the maximal change in the momentum is of the order of p itself, we find that $\Delta x > \hbar/p \geq \hbar c/E$, which coincides with the limit set by the de Broglie wavelength.

If we take this seriously—that is, a position can in principle not be measured with arbitrary accuracy—the notion of a wave function loses its meaning. On the other hand, if we would like to localise the particle more accurately than within its de Broglie wavelength, it seems to require an uncertainty in momentum that can only be achieved by allowing for negative energy states. But negative energy states will be interpreted as antiparticles, and once antiparticles are introduced, which can annihilate with particles, particle number is no longer conserved and we likewise lose the notion of position of a particle. Only a free particle, as a plane wave, seems to be compatible with relativistic invariance.

We will now verify by direct computation that localising the wave function within the light cone will indeed require negative energy states. We consider first the positive square root of the Klein–Gordon equation and solve the Schrödinger equation for the initial condition $\Psi(\vec{x}, 0) = \delta_3(\vec{x})$. From this we can solve any initial condition by convolution. As the Schrödinger equation is first order in time, the initial condition uniquely fixes the wave function for all later times, and there will be a unique answer to the question whether the wave function vanishes outside the light cone (i.e., for $t > |\vec{x}|$). Problem 1 asks you to investigate this in the simpler case of one, instead of three, spatial dimensions. For the latter we simply give the result here, using the fact that in Fourier space the solution is trivial. Computing $\Psi(\vec{x}, t)$ thus requires just some skills in performing Fourier integrals.

$$\Psi(\vec{x}, t) = \int \frac{d_3\vec{p}}{(2\pi\hbar)^3} e^{i\vec{p}\cdot\vec{x}/\hbar} e^{-it\sqrt{\vec{p}^2c^2+m^2c^4}/\hbar}$$

$$= \int \frac{p^2\, dp \sin(\theta)d\theta}{(2\pi)^2\hbar^3} e^{ipr\cos(\theta)/\hbar} e^{-it\sqrt{p^2c^2+m^2c^4}/\hbar}$$

$$= \frac{1}{2\pi^2 r\hbar^2} \int p\, dp \sin(pr/\hbar) e^{-it\sqrt{p^2c^2+m^2c^4}/\hbar}$$

$$= \frac{-i}{2\pi^2 r} \frac{\partial^2}{\partial r \partial t} \int_0^\infty dp \frac{\cos(pr/\hbar)}{\sqrt{p^2c^2+m^2c^4}} e^{-it\sqrt{p^2c^2+m^2c^4}/\hbar}. \tag{1.6}$$

We now introduce

$$p = mc\sinh(u), \quad mcr/\hbar = z\cosh(v), \quad mc^2t/\hbar = z\sinh(v),$$
$$z^2 = m^2c^2(r^2 - c^2t^2)/\hbar^2, \tag{1.7}$$

such that (the last identity simply being the definition of the modified Bessel function K_o)

$$\Psi(\vec{x}, t) = \frac{-i}{4\pi^2 rc} \frac{\partial^2}{\partial r \partial t} \int_{-\infty}^\infty du\, \cos\left(z\sinh(u)\cosh(v)\right) e^{-iz\sinh(v)\cosh(u)}$$

$$= \frac{-i}{8\pi^2 rc} \frac{\partial^2}{\partial r \partial t} \int_{-\infty}^\infty du \left(e^{-iz\sinh(u+v)} + e^{-iz\sinh(u-v)}\right)$$

$$= \frac{-i}{2\pi^2 rc} \frac{\partial^2}{\partial r \partial t} \int_0^\infty du\, \cos\left(z\sinh(u)\right) \equiv \frac{-i}{2\pi^2 rc} \frac{\partial^2}{\partial r \partial t} K_o(z). \tag{1.8}$$

Outside of the light cone, z is real ($r^2 > c^2t^2$) and $\Psi(\vec{x}, t)$ is purely imaginary. It decays exponentially, but does *not* vanish! Inside the light cone we find by analytic continuation [see, e.g., Appendix C of *Relativistic Quantum Fields* by J.D. Björken and S.D. Drell (McGraw Hill, New York, 1965)] the following explicit expression

$$\Psi(\vec{x}, t) = \frac{1}{4\pi rc} \frac{\partial^2}{\partial r \partial t} \left(i Y_o(mc\sqrt{c^2t^2 - r^2}/\hbar) - \text{sign}(t) J_o(mc\sqrt{c^2t^2 - r^2}/\hbar) \right),$$

$$r^2 < c^2 t^2. \tag{1.9}$$

If we want to insist on locality, i.e., $\Psi(\vec{x}, t) = 0$ for $|\vec{x}| > ct$, and want to stay as close as possible to the solutions of the Schrödinger equation, we could take the real part of Ψ as the wave function. It satisfies the Klein–Gordon equation but not its positive square root. Ψ^* is a solution of the negative square root of the Klein–Gordon equation and corresponds to a *negative* energy solution. Apparently, localisation is only possible if we allow for negative energy solutions.

2

Quantisation of Fields

DOI: 10.1201/b15364-2

As position is no longer a quantum observable but free particles do not seem to be in contradiction with relativistic invariance, we can try to introduce such a free particle as a quantum observable. This observable is hence described by a plane wave

$$\varphi_{\vec{k}}(\vec{x}, t) = e^{-i(k_0 t - \vec{x} \cdot \vec{k})/\hbar}, \tag{2.1}$$

which satisfies the Klein–Gordon equation

$$-\hbar^2 \frac{\partial^2 \varphi(\vec{x}, t)}{\partial t^2} = -\hbar^2 c^2 \frac{\partial^2 \varphi(\vec{x}, t)}{\partial \vec{x}^2} + m^2 c^4 \varphi(\vec{x}, t), \tag{2.2}$$

where $k_0 = \sqrt{c^2 \vec{k}^2 + m^2 c^4}$ is the energy of the free particle. By superposition of these plane waves, we can make a superposition of free particles, which is therefore described by a *field*

$$\varphi(\vec{x}, t) = (2\pi\hbar)^{-\frac{3}{2}} \int d_3 k \, \tilde{\varphi}(\vec{k}, t) e^{i\vec{k} \cdot \vec{x}/\hbar}. \tag{2.3}$$

It satisfies the Klein–Gordon equation if the Fourier components $\tilde{\varphi}(\vec{k}, t)$ satisfy the harmonic equation

$$-\hbar^2 \frac{\partial^2 \tilde{\varphi}(\vec{k}, t)}{\partial t^2} = (c^2 \vec{k}^2 + m^2 c^4) \tilde{\varphi}(\vec{k}, t) \equiv k_0^2(\vec{k}) \tilde{\varphi}(\vec{k}, t). \tag{2.4}$$

Its solutions split in positive and negative frequency components

$$\tilde{\varphi}(\vec{k}, t) = \tilde{\varphi}_+(\vec{k}) e^{-ik_0 t/\hbar} + \tilde{\varphi}_-(\vec{k}) e^{ik_0 t/\hbar}. \tag{2.5}$$

The wave function, or rather the wave functional $\Psi(\varphi)$, describes the distribution over the various free particle states. The basic dynamical variables are $\tilde{\varphi}(\vec{k})$. These play the role the coordinates used to play in ordinary quantum mechanics and will require quantisation. As they satisfy a simple harmonic equation in time, it is natural to quantise them as harmonic oscillators.

The Hamiltonian is then simply the sum of the harmonic oscillator Hamiltonian for each $\vec{k}$, with frequency $\omega(\vec{k}) \equiv k_0(\vec{k})/\hbar$.

$$i\hbar\frac{\partial \Psi(\varphi)}{\partial t} = H\Psi(\varphi) = \sum_{\vec{k}} \mathcal{H}(\vec{k})\Psi(\varphi),$$

$$\mathcal{H}(\vec{k}) = \tfrac{1}{2}|\tilde{\pi}(\vec{k})|^2 + \tfrac{1}{2}\omega(\vec{k})^2|\tilde{\varphi}(\vec{k})|^2, \quad \tilde{\pi}(\vec{k}) \equiv \frac{\hbar}{i}\frac{\partial}{\partial \tilde{\varphi}(\vec{k})}. \tag{2.6}$$

In a finite volume with periodic boundary conditions, the integral over the momenta is replaced by a sum as the momenta are in that case discrete, $\vec{k} = 2\pi\vec{n}\hbar/L$, $\vec{n} \in \mathbb{Z}^3$. Like for the harmonic oscillator, we can introduce annihilation and creation operators

$$a(\vec{k}) = \frac{1}{\sqrt{2\hbar\omega(\vec{k})}}\left(\omega(\vec{k})\tilde{\varphi}(\vec{k}) + i\tilde{\pi}(\vec{k})\right)$$

$$a^\dagger(\vec{k}) = \frac{1}{\sqrt{2\hbar\omega(\vec{k})}}\left(\omega(\vec{k})\tilde{\varphi}^*(\vec{k}) - i\tilde{\pi}^*(\vec{k})\right), \tag{2.7}$$

and express the field operator (the equivalent of the coordinates) in terms of these creation and annihilation operators. To give the field operator its time dependence, we have to invoke the Heisenberg picture, which gives $\varphi(\vec{x},t) = e^{iHt/\hbar}\varphi(\vec{x},0)e^{-iHt/\hbar}$. Using the well-known fact that $e^{iHt/\hbar}a(\vec{k})e^{-iHt/\hbar} = e^{-i\omega(\vec{k})t}a(\vec{k})$ and $e^{iHt/\hbar}a^\dagger(\vec{k})e^{-iHt/\hbar} = e^{i\omega(\vec{k})t}a^\dagger(\vec{k})$, which is a consequence of $[a(\vec{k}), H] = \hbar\omega(\vec{k})a(\vec{k})$ and $[a^\dagger(\vec{k}), H] = -\hbar\omega(\vec{k})a^\dagger(\vec{k})$, we find

$$\varphi(\vec{x},t) = L^{-\frac{3}{2}}\sum_{\vec{k}}\frac{\hbar}{\sqrt{2k_0(\vec{k})}}\left(a^\dagger(\vec{k})e^{-i(\vec{k}\cdot\vec{x}-k_0t)/\hbar} + a(\vec{k})e^{i(\vec{k}\cdot\vec{x}-k_0t)/\hbar}\right). \tag{2.8}$$

In an infinite volume we replace $L^{-\frac{3}{2}}\sum_{\vec{k}}$ by $(2\pi\hbar)^{-\frac{3}{2}}\int d_3\vec{k}$. Note that in the Heisenberg picture, positive energy modes behave in time as $e^{iEt/\hbar}$. Apparently we can identify (up to a factor) $\tilde{\varphi}_-(\vec{k})$ with $a^\dagger(-\vec{k})$ and $\tilde{\varphi}_+(\vec{k})$ with $a(\vec{k})$, which is compatible with $\tilde{\varphi}^*(\vec{k}) = \tilde{\varphi}(-\vec{k})$, required to describe a real field (complex fields will be discussed in Problem 5).

The Hilbert space is now given by the product of the Hilbert spaces of each $\vec{k}$ separately

$$|\{n_{\vec{k}}\}> = \prod_{\vec{k}}|n_{\vec{k}}> = \prod_{\vec{k}}\frac{a^\dagger(\vec{k})^{n_{\vec{k}}}}{\sqrt{n_{\vec{k}}!}}|0_{\vec{k}}>, \tag{2.9}$$

with $n_{\vec{k}}$ the occupation number, which in field theory is now interpreted as the *number* of free particles of momentum $\vec{k}$, a definition that makes sense as the energy of such a state is $n_{\vec{k}}k_0(\vec{k})$ above the state with zero occupation number (the 'vacuum'). It is the property of the harmonic oscillator that its energy is linear in the occupation number, which makes the field theory interpretation in terms of particles possible. The annihilation operator in this language therefore removes a particle (lowering the energy by the appropriate amount), which consequently can be interpreted as the annihilation of the removed particle with an antiparticle (described by the annihilation operator). For a real scalar field, a particle is its own antiparticle and this description is perhaps somewhat unfamiliar. But for the complex field of Problem 5, the Fourier component with negative energy is independent of the one with positive energy, hence describing a separate degree of freedom, namely that of an antiparticle with opposite charge.

Interactions between the particles are simply introduced by modifying the Klein–Gordon equation to have nonlinear terms, after which in general the different Fourier components no longer decouple. Field theory thus seems to be nothing but the quantum mechanics of an infinite number of degrees of freedom. It is, however, its physical interpretation that crucially differs from that of ordinary quantum mechanics. It is this interpretation that is known as *second quantisation*. We were forced to introduce the notion of fields and the interpretation involving antiparticles when combining quantum mechanics with relativistic invariance. We should therefore verify that indeed it does not give rise to propagation of information with a speed larger than the speed of light. This is implied by the following identity, which for the free scalar field will be verified in Problem 6:

$$[\varphi(\vec{x}, t), \varphi(\vec{x}', t')] = 0, \quad \text{for} \quad (\vec{x} - \vec{x}')^2 > (t - t')^2 c^2. \tag{2.10}$$

It states that the action of an operator on the wave functional at a given space-time point is independent of the action of the operator at another space-time point, as long as these two points are not causally connected. Due to the description of the time evolution with a Hamiltonian, which requires the choice of a time coordinate, it remains to be established that these equations are covariant under Lorentz transformations. We will resolve this by using the path integral approach, in which the Lorentz invariance is intrinsic but which can also be shown to be equivalent to the Hamiltonian formulation.

Before preparing for path integrals by discussing the action principle, we would first like to address a simple physical consequence of the introduction and subsequent quantisation of fields. It states that empty space (all occupation numbers equal to zero) has nevertheless a nontrivial structure, in the same way that the ground state of a hydrogen atom is nontrivial. Put differently, empty space is still full of zero-point fluctuations, which are, however, only visible if we probe that empty space in one way or another. Also, formally, as each zero-point energy is nonzero, the energy of the vacuum in field theory seems to be infinite

$$E_0 = \sum_{\vec{k}} \sqrt{\vec{k}^2 c^2 + m^2 c^4} = \cdots ? \tag{2.11}$$

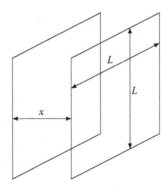

FIGURE 2.1
Vanishing field at the plates.

However (as long as gravity is left out of our considerations), one is only sensitive to differences in energy. If we probe the vacuum, its energy can *only* be put to zero for one particular value of the probe. The *dependence* of the vacuum energy on the probe can be used to discover the nontrivial structure of the vacuum.

A famous and elegant method for probing the vacuum was introduced by Casimir [*Proc. Kon. Ned. Acad. Wet.*, ser. B51 (1948) 793], who considered using two conducting plates in empty space. The energy of the vacuum is a function of the distance between the two plates, which gives a force. Strictly speaking, we should discuss this in the situation of the quantised electromagnetic field (see Itzykson and Zuber, par. 3-2-4), but the essential ingredient is that Fourier components of the field are affected by the presence of the conducting plates. We can also discuss this in the context of the simple scalar field we have introduced before, by assuming that the field has to vanish at the plates, see Figure 2.1. For simplicity we will also take the mass of the scalar particles to vanish. If furthermore we use periodic boundary conditions in the two other perpendicular directions over a distance L, then one easily verifies that the force per unit area on the conducting plates is given by

$$F_L(x) = -dE_o(x)/dx = -\frac{1}{2L^2}\frac{d}{dx}\sum_{\vec{n}\in\mathbb{Z}^2}\sum_{k=1}^{\infty}\sqrt{\left(\frac{2\pi\hbar c\vec{n}}{L}\right)^2 + \left(\frac{\pi\hbar ck}{x}\right)^2}, \quad (2.12)$$

where due to the vanishing boundary conditions the Fourier modes in the x_1 direction, perpendicular to the conducting walls, are given by $\sin(\pi k x_1/x)$ with k a positive integer, whereas the quantisation of the momenta in the other two directions is as usual.

One can now formally take the infinite volume limit

$$F(x) = \lim_{L\to\infty} F_L(x) = -\frac{c}{8\pi^2\hbar^2}\frac{d}{dx}\sum_{k=1}^{\infty}\int d_2\vec{p}\sqrt{\vec{p}^2 + \left(\frac{\pi\hbar k}{x}\right)^2}. \quad (2.13)$$

The integral and the sum are clearly divergent, but as Casimir observed, in practise no conducting plate can shield a field perfectly and, especially for high frequency the boundary conditions should be modified. One can mimic this by artificially cutting off the integral and sum at high momenta. We would not expect the physical result to depend on the details of how we do this, as otherwise we could use this experiment in an ingenious way to learn how nature behaves at arbitrarily high energies. Indeed Casimir's careful analysis showed that the result is independent of the cutoff function chosen. It is an important example of what we will later recognise as renormalisability of field theory. Since the result is insensitive to the method of regularisation [only an overall constant contribution to $E_o(x)$ depends on it, but that is not observable, as we argued before], we can choose a convenient way to perform the calculation. Details of this will be provided in Problem 2. The method of calculation is known as dimensional regularisation, where one works in an arbitrary dimension ($n \neq 2$) and then analytically extends the result to $n = 2$. We will find that

$$F(x) = \lim_{n \to 2} -\frac{c}{8\pi^2\hbar^2} \frac{d}{dx} \sum_{k=1}^{\infty} \int d_n\vec{p}\sqrt{\vec{p}^2 + \left(\frac{\pi\hbar k}{x}\right)^2}$$

$$= \lim_{n \to 2}(n + 1)\zeta(-n - 1)\pi^{3n/2}\frac{\hbar c}{8\pi}\frac{\Gamma\left(-(n + 1)/2\right)}{\Gamma(-1/2)}x^{-(n+2)}. \quad (2.14)$$

in which $\zeta(i) \equiv \sum_{k=1}^{\infty} k^{-i}$ is the Riemann ζ function. It can be analytically extended to odd negative arguments, where in terms of Bernoulli coefficients $\zeta(1 - 2i) = -B_{2i}/(2i)$. Also $\Gamma(-\frac{1}{2}) = -\frac{3}{2}\Gamma(-\frac{3}{2})$ is finite, and we simply find that

$$F(x) = -\frac{\pi^2\hbar c}{480x^4}. \quad (2.15)$$

Please note that we have disregarded the space *outside* the conducting plates. Imposing also periodic boundary conditions in that direction, one easily finds that the region outside the plates contributes with $F(L - x)$ to the force and vanishes when $L \to \infty$. Therefore, the effect of the zero-point fluctuations in the vacuum leads to a (very small) attractive force, which was ten years later experimentally measured by Sparnaay [*Physica*, 24 (1958) 751]. Another famous example of the influence of zero-point fluctuations is the Lamb shift in atomic spectra (hyperfine splittings), to be discussed at the end of Chapter 22.

3

Euler–Lagrange Equations

DOI: 10.1201/b15364-3

The Klein–Gordon equation in Lorentz covariant form $[x \equiv (ct, x, y, z) \equiv (x_0; \vec{x})]$

$$g^{\mu\nu}\partial_\mu\partial_\nu\varphi(x) + m^2\varphi(x) = 0, \quad g^{\mu\nu} = \begin{pmatrix} 1 & & & \ominus \\ & -1 & & \\ & & -1 & \\ \ominus & & & -1 \end{pmatrix}, \tag{3.1}$$

can be derived by variational calculus from an action principle

$$S = \int d_4 x \mathcal{L}(\varphi, \partial_\mu\varphi, x), \quad \mathcal{L}(\varphi, \partial_\mu\varphi, x) = \tfrac{1}{2}(\partial_\mu\varphi)^2 - V(\varphi),$$

$$(\partial_\mu\varphi)^2 \equiv \partial_\mu\varphi\partial^\mu\varphi = g^{\mu\nu}\partial_\mu\varphi\partial_\nu\varphi, \quad V(\varphi) = \tfrac{1}{2}m^2\varphi^2. \tag{3.2}$$

We assume the field to be given at the boundary of the domain M of integration (typically assuming the field vanishes at infinity) and demand the action to be stationary with respect to any variation $\varphi(x) \to \varphi(x) + \delta\varphi(x)$ of the field,

$$\delta S(\varphi) \equiv S(\varphi + \delta\varphi) - S(\varphi) = \int_M d_4 x \left(\partial^\mu\varphi\partial_\mu\delta\varphi - \frac{\partial V(\varphi)}{\partial\varphi}\delta\varphi \right)$$

$$= \int_M d_4 x \left(-\delta\varphi \left(\partial_\mu\partial^\mu\varphi + \frac{\partial V(\varphi)}{\partial\varphi} \right) \right) + \int_{\partial M} d_\mu\sigma (\delta\varphi\partial^\mu\varphi) = 0, \tag{3.3}$$

where $d_\mu\sigma$ is the integration measure on the boundary ∂M. The variation $\delta\varphi$ is arbitratry, except at ∂M, where we assume $\delta\varphi$ vanishes, and this implies the Euler–Lagrange equation

$$\partial_\mu\partial^\mu\varphi + \frac{\partial V(\varphi)}{\partial\varphi} = 0, \tag{3.4}$$

which coincides with the Klein–Gordon equation. We can also write the Euler–Lagrange equations for arbitrary action $S(\varphi)$ in terms of functional derivatives

$$\frac{\delta S}{\delta\varphi(x)} = \frac{\delta S}{\delta\varphi(x)} - \partial_\mu \frac{\delta S}{\delta\partial_\mu\varphi(x)} = 0, \tag{3.5}$$

where δ stands for the total functional derivative, which is then split according to the explicit dependence of the action on the field and its derivatives (usually an action will not contain higher than first-order space-time derivatives). Please note that a functional derivative has the property $\delta\varphi(x)/\delta\varphi(y) = \delta_4(x - y)$, which is why in the above equation we take functional derivatives of the action S and not, as one sees often, of the Lagrangian density $\mathcal{L}$.

The big advantage of using an action principle is that S is a Lorentz scalar, which makes it much easier to guarantee Lorentz covariance. As the action will be the starting point of the path integral formulation of field theory, Lorentz covariance is much easier to establish within this framework. (There are instances where the regularisation, required to make sense of the path integral, destroys the Lorentz invariance, like in string theory. Examples of these *anomalies* will be discussed later for the breaking of scale invariance and gauge invariance.) It is now simple to add interactions to the Klein–Gordon equation by generalising the dependence of the 'potential' $V(\varphi)$ to include higher-order terms, like

$$V(\varphi) = \tfrac{1}{2}m^2\varphi^2 + \frac{g}{4!}\varphi^4, \tag{3.6}$$

which is known as a scalar φ^4 field theory. Later we will see that one cannot add arbitrary powers of the field to this potential, except in two dimensions.

As in classical field theory, we can derive from a Lagrangian with $\varphi(x)$ and $\dot{\varphi}(x) \equiv \partial\varphi(x)/\partial t$ as its independent variables, the Hamiltonian through a Legendre transformation to the canonical pair of variables $\pi(x)$ (the 'momentum') and $\varphi(x)$ (the 'coordinate')

$$\pi(x) = \frac{\delta S}{\delta\dot{\varphi}(x)}, \quad H = \int \mathcal{H}(x)d_3\vec{x} = \int \big(\pi(x)\dot{\varphi}(x) - \mathcal{L}(x)\big)d_3\vec{x}. \tag{3.7}$$

The classical Hamilton equations of motion are given by

$$\dot{\varphi}(x) = \frac{\delta H}{\delta\pi(x)}, \quad \dot{\pi}(x) = -\frac{\delta H}{\delta\varphi(x)} + \partial_i\frac{\delta H}{\delta\partial_i\varphi(x)}. \tag{3.8}$$

For the Klein–Gordon field we simply find

$$\mathcal{H} = \tfrac{1}{2}\pi(x)^2 + \tfrac{1}{2}(\partial_i\varphi(x))^2 + \tfrac{1}{2}m^2\varphi^2(x), \tag{3.9}$$

and in Problem 5 one will see that this Hamiltonian coincides with Equation (2.6), if we substitute for $\varphi(x)$ Equation (2.8). For an interacting scalar field one finds

$$\mathcal{H} = \tfrac{1}{2}\pi(x)^2 + \tfrac{1}{2}\big(\partial_i\varphi(x)\big)^2 + V\big(\varphi(x)\big), \tag{3.10}$$

which perhaps explains why V is called the potential.

It is well known that the Hamiltonian equations imply that H itself is conserved with time, provided the Lagrangian (or Hamiltonian) has no explicit time dependence

$$\frac{dH}{dt} = \int d_3\vec{x} \left(\dot{\pi}(x) \frac{\delta H}{\delta \pi(x)} + \dot{\varphi}(x) \frac{\delta H}{\delta \varphi(x)} \right) = 0. \tag{3.11}$$

Conservation of energy is one of the most important laws of nature, and it is instructive to derive it more directly from the fact that $\mathcal{L}$ does not depend explicitly on time. We define the Lagrangian L as an integral of the Lagrange density $\mathcal{L}$ over space, $L \equiv \int d_3\vec{x}\mathcal{L}$, such that

$$\begin{aligned}
\frac{dL}{dt} &= \int d_3\vec{x} \left(\partial_t\varphi(x) \frac{\delta S}{\delta\varphi(x)} + \partial_t\left(\partial_\mu\varphi(x)\right) \frac{\delta S}{\delta\partial_\mu\varphi(x)} \right) \\
&= \int d_3\vec{x}\, \partial_\mu \left(\partial_t\varphi(x) \frac{\delta S}{\delta\partial_\mu\varphi(x)} \right).
\end{aligned} \tag{3.12}$$

The last term contains a total derivative, which vanishes if we assume that the field is time independent (or vanishes) at the boundary of the spatial integration domain. The above equation becomes now

$$\frac{dL}{dt} = \frac{d}{dt} \int d_3\vec{x}\, \dot{\varphi}(x) \frac{\delta S}{\delta\varphi(x)} = \frac{d}{dt} \int d_3\vec{x}\, \dot{\varphi}(x)\pi(x), \tag{3.13}$$

which can also be expressed as

$$\frac{d}{dt} \int d_3\vec{x} \left(\dot{\varphi}(x)\pi(x) - \mathcal{L} \right) \equiv \frac{dH}{dt} = 0. \tag{3.14}$$

In the same fashion one proves conservation of momentum in case the Lagrangian does not explicitly depend on space ($\partial\mathcal{L}/\partial x_i = 0$)

$$0 = \frac{dL}{dx_i} = \int d_3\vec{x}\, \partial_\mu \left(\partial_i\varphi(x) \frac{\delta S}{\delta\partial_\mu\varphi(x)} \right) = \frac{d}{dt} \int d_3\vec{x}\, \pi(x)\partial_i\varphi(x). \tag{3.15}$$

The conserved momentum is hence given by

$$P_i = \int d_3\vec{x}\, \pi(x)\partial_i\varphi(x). \tag{3.16}$$

Both conservation of momentum and energy are examples of conservation laws that are consequence of symmetries (translation and time invariance). They can be derived as the space integral of the time component of a conserved current or tensor

$$\partial_\mu J^\mu(x) = 0, \quad \partial_\mu T^{\mu\nu}(x) = 0. \tag{3.17}$$

In Problem 3 these quantities will be defined for a charged scalar field, where $J_\mu(x)$ can be identified with the current, whose time component is the charge density. Indeed the total charge is conserved. Assuming the current to vanish at spatial infinity, one easily finds

$$\frac{d}{dt} \int d_3\vec{x}\, J_0(x) = \int d_3\vec{x}\, \partial_i J_i(x) = 0. \tag{3.18}$$

The underlying principle is described by the Noether theorem, which implies that if the Lagrangian $\mathcal{L}$ is invariant under $\varphi \to \varphi_\Lambda$, where Λ is a parameter (such as a shift of the coordinates or a phase rotation of a complex field), then the following current is conserved:

$$J^\mu(x) = \frac{\delta S}{\delta\big(\partial_\mu\varphi(x)\big)}\frac{\partial\varphi_\Lambda(x)}{\partial\Lambda}. \tag{3.19}$$

The proof is simple and uses the Euler–Lagrange equations to substitute $\partial_\mu\{\delta S/\delta[\partial_\mu\varphi(x)]\}$ for $\delta S/\delta\varphi(x)$

$$0 = \frac{d\mathcal{L}(\varphi_\Lambda)}{d\Lambda} = \frac{\delta S}{\delta\varphi(x)}\frac{\partial\varphi_\Lambda(x)}{\partial\Lambda} + \frac{\delta S}{\delta\big(\partial_\mu\varphi(x)\big)}\frac{\partial\big(\partial_\mu\varphi_\Lambda(x)\big)}{\partial\Lambda}$$

$$= \partial_\mu\left(\frac{\delta S}{\delta\big(\partial_\mu\varphi(x)\big)}\right)\frac{\partial\varphi_\Lambda(x)}{\partial\Lambda} + \frac{\delta S}{\delta\big(\partial_\mu\varphi(x)\big)}\frac{\partial\big(\partial_\mu\varphi_\Lambda(x)\big)}{\partial\Lambda} = \partial_\mu J^\mu(x). \tag{3.20}$$

We here considered the invariance under a *global* symmetry, but important in nature are also the *local* symmetries, like the gauge invariance related to local changes of phase and the general coordinate invariance in general relativity. Particularly with the latter in mind, we demand therefore that the action S (and not just $\mathcal{L}$) is invariant under $\varphi(x) \to \varphi_{\Lambda(x)}(x)$, with Λ an arbitrary function of space-time. This actually leads to the same conserved currents in case $\mathcal{L}$ is also invariant. The same computation as above, *still using the Euler–Lagrange equations*, shows that

$$0 = \frac{\delta S}{\delta\Lambda(x)} = \partial_\mu J^\mu(x). \tag{3.21}$$

As an important example, we will discuss how this construction leads to conservation of the energy-momentum tensor, using general coordinate invariance, which is the local version of translation invariance. For this we have to make the action invariant under such local coordinate redefinitions. As long as indices are contracted with the metric tensor g, $\mathcal{L}$ will be invariant under general coordinate transformations, due to the transformation property

$$\bar{x}^\mu = x^\mu + \varepsilon^\mu(x), \quad \bar{g}^{\mu\nu}(\bar{x}) = \frac{\partial\bar{x}^\mu}{\partial x^\alpha}\frac{\partial\bar{x}^\nu}{\partial x^\beta}g^{\alpha\beta}(x). \tag{3.22}$$

For global translation invariance, ε^μ is constant, and equations (3.14) and (3.15) can be easily generalised to show that the energy-momentum tensor, $T_{\mu\nu} = \partial_\mu\varphi\partial_\nu\varphi - g_{\mu\nu}\mathcal{L}$, is conserved [Equation (3.17)]. For ε^μ not constant, we note that the integration measure d_4x is not a scalar under general coordinate transformations, but the associated Jacobian can be easily absorbed by $\sqrt{-\det g}$, where the determinant is applied to the 4×4 matrix $g_{\mu\nu}$. For a scalar field this leads to the following invariant action:

$$S = \int d_4 x \sqrt{-\det g} \left(\tfrac{1}{2} g^{\mu\nu} \partial_\mu \varphi \partial_\nu \varphi - V(\varphi) \right). \tag{3.23}$$

For the original coordinates x of Minkowski space-time, the metric is given as in Equation (3.1), in particular $\sqrt{-\det g(x)} = 1$, and by expanding $\bar{g}$ to first order in $\varepsilon^\mu(x)$ we find

$$S = \int d_4 \bar{x} \left(1 - \partial_\lambda \varepsilon^\lambda(x) \right) \left[\tfrac{1}{2} g^{\mu\nu} \partial_\mu \varphi(\bar{x}) \partial_\nu \varphi(\bar{x}) - V(\varphi(\bar{x})) \right]$$
$$+ \partial_\alpha \varepsilon^\mu(x) \partial_\mu \varphi(\bar{x}) \partial^\alpha \varphi(\bar{x}). \tag{3.24}$$

Now observe that $g^{\mu\nu}(x)$ is constant, such that the independent term of ε is a function of $\bar{x}$, integrated over $\bar{x}$, which is simply the action itself, as $\bar{x}$ now plays the role of a dummy integration variable. The linear term in ε therefore has to vanish, but note that it *only* involved the variation of the metric under the general coordinate transformation. Hence,

$$0 = \int d_4 x \, \varepsilon_\mu(x) \partial_\alpha \left(g^{\mu\alpha} \mathcal{L}(x) - \partial^\mu \varphi(x) \partial^\alpha \varphi(x) \right)$$
$$\equiv - \int d_4 x \, \varepsilon_\mu(x) \partial_\alpha T^{\alpha\mu}(x), \tag{3.25}$$

which implies conservation of the energy-momentum tensor ($T_{oo} = \mathcal{H}$). From the fact that $\delta g_{\mu\nu} = -\partial_\mu \varepsilon_\nu - \partial_\nu \varepsilon_\mu$, $\delta g^{\mu\nu} = -g^{\mu\alpha} \delta g_{\alpha\beta} g^{\beta\nu}$ and $\delta \sqrt{-\det g} = \tfrac{1}{2} g^{\mu\nu} \sqrt{-\det g} \delta g_{\mu\nu}$, we derive the identity

$$T^{\mu\nu}(x) \equiv -2 \frac{\delta S}{\delta g_{\mu\nu}(x)}. \tag{3.26}$$

In taking the derivative with respect to the metric, it is important that any Lorentz vector (like the derivative $\partial_\mu \varphi$) or tensor appears in the Lagrangian with its indices down. Furthermore, the result is to be evaluated for Minkowski space. Equation (3.26) always gives a symmetric energy-momentum tensor and from the derivation it is clear that the result holds not only for a simple scalar field, but for any other bosonic field theory (fermions form an exception, see Problem 23) like the one for the electromagnetic field, which we discuss now.

The field is given by the tensor $F_{\mu\nu}(x)$, with $E^i(x) = -F^{0i}(x)$ its electric and $B^i(x) = -\tfrac{1}{2} \varepsilon_{ijk} F^{jk}(x)$ its magnetic components. In terms of the vector potential $A_\mu(x)$, one has

$$F_{\mu\nu}(x) = \partial_\mu A_\nu(x) - \partial_\nu A_\mu(x). \tag{3.27}$$

This already implies one of the Maxwell equations (through the so-called Jacobi or integrability conditions)

$$\partial_\mu F_{\nu\lambda} + \partial_\nu F_{\lambda\mu} + \partial_\lambda F_{\mu\nu} = 0. \tag{3.28}$$

Written as $\varepsilon^{\mu\nu\lambda\sigma} \partial_\nu F_{\lambda\sigma} = 0$, they are easily seen (resp. for $\mu = 0$ and $\mu = i$) to give

$$\mathrm{div}\,\vec{B} = 0, \quad \partial_0 \vec{B} + \mathrm{rot}\,\vec{E} = \vec{0}. \tag{3.29}$$

The dynamical equations determining the fields in terms of the currents, or the sources, $J^\mu = (c\rho; \vec{J})$ are given by

$$\partial_\mu F^{\mu\nu} = \frac{1}{c} J^\nu \quad \text{or} \quad \mathrm{div}\,\vec{E} = \rho, \quad \mathrm{rot}\,\vec{B} - \partial_0 \vec{E} = \vec{J}. \tag{3.30}$$

We have chosen Heaviside–Lorentz units and in the future we will also often choose units such that $\hbar = c = 1$.

These Maxwell equations follow from the following action:

$$S_{\mathrm{em}}(J) = \int d_4 x \left(-\tfrac{1}{4} F_{\mu\nu}(x) F^{\mu\nu}(x) - A_\mu(x) J^\mu(x) \right). \tag{3.31}$$

We note, as is well known, that the equations of motion imply that the current is conserved. With Noether's theorem this makes us suspect that this is caused by a symmetry and indeed it is known that under the gauge transformation

$$A_\mu(x) \rightarrow A_\mu(x) + \partial_\mu \Lambda(x) \tag{3.32}$$

the theory does not change. Our action is invariant under this symmetry if and only if the current is conserved. This gauge symmetry will play a crucial role in the quantisation of the electromagnetic field.

An example of a conserved current can be defined for a complex scalar field. Its action for a free particle is given by

$$S_0 = \int d_4 x \left(\partial_\mu \varphi^*(x) \partial^\mu \varphi(x) - m^2 \varphi^*(x) \varphi(x) \right). \tag{3.33}$$

It is invariant under a phase rotation $\varphi(x) \rightarrow \exp(ie\Lambda)\varphi(x)$ and from Noether's theorem we deduce that

$$J_\mu(x) \equiv ie(\varphi(x)\partial_\mu \varphi^*(x) - \varphi^*(x)\partial_\mu \varphi(x)) \tag{3.34}$$

is conserved; see Problem 3. We can extend this global phase symmetry to a local symmetry if we couple the scalar field *minimally* to the vector potential

$$S = \int d_4 x \left(-\tfrac{1}{4} F_{\mu\nu}(x) F^{\mu\nu}(x) + (D_\mu \varphi)^*(x) D^\mu \varphi(x) - m^2 \varphi^*(x)\varphi(x) \right),$$
$$D_\mu \varphi(x) \equiv \partial_\mu \varphi(x) - ie A_\mu(x)\varphi(x). \tag{3.35}$$

This guarantees the combined invariance under a local gauge transformation

$$\varphi(x) \rightarrow \exp(ie\Lambda(x))\varphi(x), \quad A_\mu(x) \rightarrow A_\mu(x) + \partial_\mu\Lambda(x), \tag{3.36}$$

which makes the *covariant* derivative $D_\mu\varphi(x)$ of the scalar field transform as the scalar field itself, even for local phase rotations. Note that we can write this action also as

$$S = S_{em}(J) + S_0 + \int d_4 x\, e^2 A_\mu(x) A^\mu(x) |\varphi(x)|^2, \tag{3.37}$$

with J as given in Equation (3.34). We leave it as an exercise to show how the action of the electromagnetic field can be generalised to be invariant under general coordinate transformations and to derive from this the energy-momentum tensor. The result is given by

$$S_{em}(J = 0) = -\tfrac{1}{4} \int d_4 x\, g^{\mu\lambda} g^{\nu\sigma} F_{\mu\nu} F_{\lambda\sigma} \sqrt{-\det g},$$
$$T^{\mu\nu} = \tfrac{1}{4} g^{\mu\nu} F^{\lambda\sigma} F_{\lambda\sigma} - F^{\mu\lambda} F^\nu{}_\lambda. \tag{3.38}$$

4

Tree-Level Diagrams

DOI: 10.1201/b15364-4

In general, in the presence of interactions, the equations of motions cannot be solved exactly, and one has to resort to a perturbative expansion in a small parameter. We discuss the scalar case first, as it is as always the simplest. We add to the Lagrangian density $\mathcal{L}$ a so-called source term, which couples linearly to the field φ (compare the driving force term for a harmonic oscillator)

$$\mathcal{L} = \tfrac{1}{2}(\partial_\mu \varphi)^2 - V(\varphi) - J(x)\varphi(x). \tag{4.1}$$

For sake of explicitness, we will take the following expression for the potential

$$V(\varphi) = \tfrac{1}{2}m^2\varphi^2(x) + \frac{g}{3!}\varphi^3(x). \tag{4.2}$$

The Euler–Lagrange equations are now given by

$$\partial_\mu \partial^\mu \varphi(x) + m^2\varphi(x) + \tfrac{1}{2}g\varphi^2(x) + J(x) = 0. \tag{4.3}$$

If $g = 0$ it is easy to solve the equation (describing a free particle interacting with a given source) in Fourier space. Introducing the Fourier coefficients

$$\tilde{J}(k) = \frac{1}{(2\pi)^2} \int d_4x \, e^{ikx} J(x), \quad \tilde{\varphi}(k) = \frac{1}{(2\pi)^2} \int d_4x \, e^{ikx} \varphi(x), \tag{4.4}$$

it follows that

$$(-k^2 + m^2)\tilde{\varphi}(k) + \tilde{J}(k) = 0 \quad \text{or} \quad \varphi(x) = \int d_4y \, G(x-y)J(y), \tag{4.5}$$

where G is called the Green's function, as it is the solution of the equation

$$(\partial_\mu \partial^\mu + m^2)G(x-y) = -\delta_4(x-y). \tag{4.6}$$

Explicitly, it is given by the following Fourier integral

$$G(x-y) = \int \frac{d_4k}{(2\pi)^4} \frac{e^{-ik(x-y)}}{k^2 - m^2 + i\varepsilon}. \tag{4.7}$$

Please note our shorthand notation of k^2 for $k_\mu k^\mu$ and $k(x-y)$ for $k_\mu(x^\mu - y^\mu)$. A Green's function is not uniquely specified by its second-order equations but also requires boundary conditions. These boundary conditions are, as we will see, specified by the term $i\varepsilon$. Because of the interpretation of the negative energy states as antiparticles, which travel 'backwards' in time, the quantum theory will require that the positive energy part vanishes for past infinity, whereas the negative energy part will be required to vanish for future infinity. Classically this would not make sense, and we would require the solution to vanish outside the future light cone. The effect of the $i\varepsilon$ prescription is to shift the poles on the real axes to the complex k_0 plane at $k_0 = \pm[(\vec{k}^2 + m^2)^{\frac{1}{2}} - i\varepsilon]$. In Chapter 5 we will see that this will imply the appropriate behaviour required by the quantum theory.

Now that we have found the solution for the free field coupled to a source, we can do perturbation in the strength of the *coupling constant g*.

$$\partial_\mu \partial^\mu \varphi(x) + m^2 \varphi(x) + J(x) = -\tfrac{1}{2}g\varphi^2(x) \tag{4.8}$$

can be solved iteratively by substituting a series expansion for $\varphi(x)$,

$$\varphi(x) = \varphi_0(x) + g\varphi_1(x) + g^2\varphi_2(x) + \cdots . \tag{4.9}$$

Obviously we have

$$\varphi_0(x) = \int d_4y \, G(x - y)J(y), \tag{4.10}$$

whereas $\varphi_1(x)$ will be determined by the equation

$$\partial_\mu \partial^\mu \varphi_1(x) + m^2 \varphi_1(x) = -\tfrac{1}{2}\varphi_0^2(x). \tag{4.11}$$

We can therefore interpret the right-hand side as a source (up to a minus sign) and this allows us to solve $\varphi_1(x)$ using the Green's function

$$\varphi_1(x) = \tfrac{1}{2} \int d_4y \, G(x - y)\varphi_0^2(y)$$

$$= \tfrac{1}{2} \int G(x - y)d_4y \int G(y - z)G(y - w)J(z)J(w)d_4z d_4w. \tag{4.12}$$

This looks particularly simple in Fourier space

$$\tilde{\varphi}_1(k) = \frac{1}{2(2\pi)^2} \frac{1}{k^2 - m^2 + i\varepsilon} \int d_4p \frac{\tilde{J}(p)\tilde{J}(k - p)}{(p^2 - m^2 + i\varepsilon)((k - p)^2 - m^2 + i\varepsilon)}. \tag{4.13}$$

It is clear that this can be continued iteratively, e.g.,

$$\partial_\mu \partial^\mu \varphi_n(x) + m^2 \varphi_n(x) = -\tfrac{1}{2} \sum_{i=0}^{n-1} \varphi_i(x)\varphi_{n-1-i}(x), \tag{4.14}$$

which is solved by

$$\varphi_n(x) = \tfrac{1}{2} \int d_4 y \, G(x - y) \big(2\varphi_0(y)\varphi_{n-1}(y) + 2\varphi_1(y)\varphi_{n-2}(y) + \cdots \big). \quad (4.15)$$

Here we have written out the terms in the sum explicitly to indicate that all terms occur twice and are the product of two different terms, except for the term $\varphi_{\frac{1}{2}(n-1)}(y)^2$ at n odd, which occurs once. In Fourier space one finds

$$\tilde{\varphi}_n(k) = \frac{1}{2(2\pi)^2} \frac{1}{k^2 - m^2 + i\varepsilon}$$

$$\times \int d_4 p \big(2\tilde{\varphi}_0(p)\tilde{\varphi}_{n-1}(k - p) + 2\tilde{\varphi}_1(p)\tilde{\varphi}_{n-2}(k - p) + \cdots \big). \quad (4.16)$$

By induction it is now easy to prove that

$$= \sum_{\text{diagrams}} \frac{g^{\#\text{vertices}}}{N(\text{diagram})} \int \prod_{\{i_v\}} dx_{i_v} \prod_{<i,j>} G(x_i - x_j) \prod_{\{k_s\}} J(x_{k_s}). \quad (4.17)$$

Here the index i_v runs over all vertices and sources (so that it does *not* label the four space-time components of a single point, frequently it will be assumed that it is clear from the context what is meant), whereas k_s runs only over positions of the sources. The expression $< i, j >$ stands for the pairs of points in a diagram connected by a line (called *propagator*).

The Feynman rules to convert a diagram to the solution are apparently that each line (propagator) between points x and y contributes $G(x - y)$ and each cross (source) at a point x contributes $J(x)$. Furthermore, for each vertex at a point x we insert $\int d_4 x$ and a power of the coupling constant g. Finally each diagram comes with an overall factor $1/N(\text{diagram})$, being the inverse of the order of the permutation group (interchange of lines and vertices) that leaves the diagram invariant (which is also the number of ways the diagram can be constructed out of its building blocks). We have derived these rules for the case that $\lambda = 0$, such that only three-point vertices appear. All that is required to generalise this to the arbitrary case with n-point vertices is that each of these comes with its own coupling constant (i.e., λ for a four-point vertex). This is the reason why these vertices are weighed by a factor $1/n!$ in the potential and hence by a factor $1/(n-1)!$ in the equations of motion. [To be precise, if $V(\varphi) = g_n \varphi^n / n!$, the equation of motion gives $\partial_\mu^2 \varphi(x) + J(x) = -g_n \varphi^{(n-1)}(x)/(n - 1)!$, and the factor $(n - 1)!$ is part of the combinatorics involved in interchanging each of the $n - 1$ factors φ in the interaction term.]

TABLE 4.1

Feynman rules.

Coordinate space		Momentum space	
$\equiv g \int d_4 x$		$\equiv \dfrac{g}{(2\pi)^2}\delta_4(\sum_i k_i)$	vertex
$\equiv G(x-y)$		$\equiv \int d_4 k \dfrac{1}{k^2 - m^2 + i\varepsilon}$	propagator
$\equiv \int d_4 x J(x)$		$\equiv \tilde{J}(k)$	source

It is straightforward to translate these Feynman rules to momentum space, by inserting the Fourier expansion of each of the terms that occur. Each propagator which carries a momentum k is replaced by a factor $1/(k^2-m^2+i\varepsilon)$ and $\int d_4 k$, each source with momentum k flowing in the source by a factor $\tilde{J}(k)$, each vertex by a factor of the coupling constant (i.e., g_n for an n-point function), a factor $1/(2\pi)^2$ [for an n-point function a factor $(2\pi)^{4-2n}$] and a momentum conserving delta function, see Table 4.1. To understand why momentum is conserved at each vertex we use that in the coordinate formulation each vertex comes with an integration over its position. As each line entering the vertex carries a Green's function that depends on that position (this being the only dependence), we see that a vertex at point x gives rise to

$$\int d_4 x \prod_\alpha G(x - x_\alpha) \rightarrow \int d_4 x \prod_\alpha \int \frac{d_4 k_\alpha \, e^{-ik_\alpha(x-x_\alpha)}}{(2\pi)^4(k_\alpha^2 - m^2 + i\varepsilon)}$$

$$= (2\pi)^4 \prod_\alpha \int \frac{d_4 k_\alpha \, e^{ik_\alpha x_\alpha}}{(2\pi)^4(k_\alpha^2 - m^2 + i\varepsilon)} \delta_4\left(\sum_\alpha k_\alpha\right). \quad (4.18)$$

Conventions in the literature can differ on how the factors of i (which will appear in the quantum theory) and 2π are distributed over the vertices and propagators. Needless to say, the final answers have to be independent of the chosen conventions.

As a last example in this section, we will look again at the electromagnetic field (whose particles are called photons). In Fourier space the equations of motion are given by

$$(-k^2\delta_\mu^\nu + k_\mu k^\nu)\tilde{A}^\mu(k) = \tilde{J}^\nu(k). \quad (4.19)$$

Unfortunately the matrix $-k^2\delta_\mu^\nu + k_\mu k^\nu$ has no inverse as k^μ is an eigenvector with zero eigenvalue. This is a direct consequence of the gauge invariance as the gauge transformation of Equation (3.32) in Fourier language reads

$$\tilde{A}_\mu(k) \rightarrow \tilde{A}_\mu(k) + ik_\mu \tilde{\Lambda}(k). \quad (4.20)$$

The component of A_μ in the direction of k_μ is for obvious reasons called the longitudinal component, which can be fixed to a particular value by a gauge transformation. Fixing the longitudinal component of the electromagnetic field (also called *photon field*) is called gauge fixing, and the gauge choice is prescribed by the gauge condition. An important example is the so-called *Lorentz gauge*

$$\partial_\mu A^\mu(x) = 0 \quad \text{or} \quad k_\mu \tilde{A}^\mu(k) = 0. \tag{4.21}$$

Because of the gauge invariance, the choice of gauge has no effect on the equations of motion because the current is conserved, or $k^\mu \tilde{J}_\mu(k) = 0$. The current (i.e., the source) does not couple to the unphysical longitudinal component of the photon field. It stresses again the importance of gauge invariance and its associated conservation of currents.

To impose the gauge fixing, we can add a term to the Lagrangian which enforces the gauge condition. Without such a term the action is stationary under any longitudinal variation $\delta A_\mu(x) = \partial_\mu \Lambda(x)$ of the vector field, and the added term should be such that stationarity in that direction imposes the gauge condition. For any choice of the parameter $\alpha \neq 0$ this is achieved by the action

$$S = \int d_4 x \left(-\tfrac{1}{4} F_{\mu\nu}(x) F^{\mu\nu}(x) - \tfrac{1}{2}\alpha \big(\partial_\mu A^\mu(x)\big)^2 - A_\mu(x) J^\mu(x) \right). \tag{4.22}$$

Indeed, the variation $\delta A_\mu(x) = \partial_\mu \Lambda(x)$ in the longitudinal direction leads to the equation

$$-\alpha \int d_4 x \, \partial_\mu \partial^\mu \Lambda(x) \partial_\nu A^\nu(x) = 0, \tag{4.23}$$

which implies the Lorentz gauge (assuming vanishing boundary conditions for the vector potential at infinity).

The equations of motion for this action now yield

$$\partial_\mu F^{\mu\nu}(x) + \alpha \partial^\nu \partial_\mu A^\mu(x) = J^\nu(x), \tag{4.24}$$

or in Fourier space

$$\big(-k^2 \delta^\nu_\mu + (1-\alpha)k_\mu k^\nu \big) \tilde{A}^\mu(k) = \tilde{J}^\nu(k), \tag{4.25}$$

which is invertible, as long as $\alpha \neq 0$. The result is given by

$$\tilde{A}^\mu(k) = -\frac{\delta^\mu_\nu - \frac{k^\mu k_\nu(1-\alpha^{-1})}{k^2 + i\varepsilon}}{k^2 + i\varepsilon} \tilde{J}^\nu(k). \tag{4.26}$$

This is consequently the propagator of the electromagnetic field (in the Lorentz gauge), also simply called the photon propagator. Like in the scalar case, it can be used to perform a perturbative expansion for the classical equations of motion.

Note that the photon propagator simplifies dramatically if we choose $\alpha = 1$, but all final results should be independent of the choice of α and even of the choice of gauge fixing all together. This is the hard part in gauge theories. One needs to fix the gauge to perform perturbation theory and then one has to prove that the result does not depend on the choice of gauge fixing. In quantum theory this is not entirely trivial, as the regularisation can break the gauge invariance explicitly. Fortunately, there are regularisations that preserve the gauge invariance, like dimensional regularisation, which we already encountered in Chapter 2 (in discussing the Casimir effect). In the presence of fermions, the situation can, however, be considerably more tricky. Some different choices of gauge fixing will be explored in Problems 8 and 9.

5

Hamiltonian Perturbation Theory

DOI: 10.1201/b15364-5

We consider the Hamiltonian for a free scalar particle coupled to a source. We will see that the source can be used to create particles from the vacuum in quantum theory, and it forms an important ingredient, like for the derivation of the classical perturbation theory of the previous chapter, in deriving scattering amplitudes and cross sections. Also the Green's function will reappear, but now with a unique specification of the required boundary conditions following from the time ordering in the quantum evolution equations.

For the Lagrangian

$$\mathcal{L} = \tfrac{1}{2}\partial_\mu \varphi \partial^\mu \varphi - \tfrac{1}{2}m^2\varphi^2 - \bar{\varepsilon} J\,\varphi, \tag{5.1}$$

the Hamiltonian is given by

$$\mathcal{H} = \tfrac{1}{2}\pi^2 + \tfrac{1}{2}(\partial_i \varphi)^2 + \tfrac{1}{2}m^2\varphi^2 + \bar{\varepsilon} J\,\varphi, \tag{5.2}$$

where $\bar{\varepsilon}$ is a small expansion parameter. We will quantise the theory in a finite volume $V = [0, L]^3$ with periodic boundary conditions, such that the momenta are discrete, $\vec{k} = 2\pi\vec{n}/L$.

$$\varphi(\vec{x}, t=0) = \sum_{\vec{k}} \frac{1}{\sqrt{2Vk_o(\vec{k})}} \left(a(\vec{k})e^{i\vec{k}\cdot\vec{x}} + a^\dagger(\vec{k})e^{-i\vec{k}\cdot\vec{x}} \right),$$

$$\pi(\vec{x}, t=0) = -i \sum_{\vec{k}} \sqrt{\frac{k_o(\vec{k})}{2V}} \left(a(\vec{k})e^{i\vec{k}\cdot\vec{x}} - a^\dagger(\vec{k})e^{-i\vec{k}\cdot\vec{x}} \right). \tag{5.3}$$

The Hamiltonian is now given by $H(t) = H_0 + \bar{\varepsilon} H_1(t)$, and we work out the perturbation theory in the Schrödinger representation. We have

$$H_0 = \sum_{\vec{k}} k_0(\vec{k})(a^\dagger(\vec{k})a(\vec{k}) + \tfrac{1}{2}),$$

$$H_1(t) = \int_V d_3\vec{x}\, J(\vec{x}, t)\varphi(\vec{x}) = \sum_{\vec{k}} \frac{1}{\sqrt{2k_0(\vec{k})}} J(\vec{k}, t) \left(a(-\vec{k}) + a^\dagger(\vec{k}) \right), \tag{5.4}$$

here $\tilde{J}(\vec{k}, t)$ is the Fourier coefficient of $J(\vec{x}, t)$, or

$$J(\vec{x}, t) = \frac{1}{\sqrt{V}} \sum_{\vec{k}} \tilde{J}(\vec{k}, t) e^{i\vec{k}\cdot\vec{x}}. \tag{5.5}$$

Let us start at $t = 0$ with the vacuum state $|0>$, which has the property that $a(\vec{k})|0> = 0$ for all momenta, then it follows that

$$\frac{d}{dt}|\Psi(t)> = -iH(t)|\Psi(t)>, \tag{5.6}$$

which can be evaluated by perturbing in $\bar{\varepsilon}$.

$$|\Psi(t)> \equiv e^{-iH_0 t}|\hat{\Psi}(t)>, \quad |\hat{\Psi}(t)> = \sum_{n=0}^{\infty} \bar{\varepsilon}^n |\hat{\Psi}_n(t)>,$$

$$\frac{d}{dt}|\hat{\Psi}_n(t)> = -ie^{iH_0 t} H_1(t) e^{-iH_0 t}|\hat{\Psi}_{n-1}(t)>. \tag{5.7}$$

Actually, by transforming to $|\hat{\Psi}(t)>$ we are using the interaction picture, which is the usual way of performing Hamiltonian perturbation theory known from ordinary quantum mechanics. These equations can be solved iteratively as follows

$$|\hat{\Psi}_1(t)> = -i \int_0^t dt_1\, e^{iH_0 t_1} H_1(t_1) e^{-iH_0 t_1}|0>,$$

$$|\hat{\Psi}_2(t)> = -i \int_0^t dt_1\, e^{iH_0 t_1} H_1(t_1) e^{-iH_0 t_1}|\hat{\Psi}_1(t_1)>$$

$$= -\int_0^t dt_1\, e^{iH_0 t_1} H_1(t_1) \int_0^{t_1} dt_2\, e^{iH_0(t_2-t_1)} H_1(t_2) e^{-iH_0 t_2}|0>,$$

$$|\hat{\Psi}_n(t)> = -i \int_0^t dt_1\, e^{iH_0 t_1} H_1(t_1) e^{-iH_0 t_1}|\hat{\Psi}_{n-1}(t_1)> = \cdots. \tag{5.8}$$

Please note the time ordering, which is essential as $H_1(t)$ does not commute with $H_1(t')$ for different t and t'. We can, for example, compute the probability that at time t $|\Psi(t)>$ is still in the ground state (whose energy we denote by E_0, which will often be assumed to vanish)

$$< 0|\Psi(t)> = e^{-iE_0 t} < 0|\hat{\Psi}(t)> = e^{-iE_0 t} \left\{ 1 - i\bar{\varepsilon} \int_0^t dt_1\, < 0|H_1(t_1)|0> \right.$$

$$\left. - \bar{\varepsilon}^2 \int_0^t dt_1 \int_0^{t_1} dt_2\, < 0|H_1(t_1) e^{i(H_0-E_0)(t_2-t_1)} H_1(t_2)|0> + \mathcal{O}(\bar{\varepsilon}^3) \right\}. \tag{5.9}$$

It is simple to see that the term linear in $\bar{\varepsilon}$ will vanish, as the vacuum expectation values of the creation and annihilation operators vanish, i.e.,

$< 0|a^{\dagger}|0 >=< 0|a|0 >= 0$. To evaluate the remaining expectation value in the above equation, we substitute H_1 in terms of the creation and annihilation operators [see Equation (5.4)]

$$< 0|H_1(t_1)e^{i(H_0-E_0)(t_2-t_1)}H_1(t_2)|0 > = \sum_{\vec{k},\vec{p}} \frac{\tilde{J}(\vec{p},t_1)\tilde{J}(\vec{k},t_2)}{\sqrt{4k_0(\vec{p})k_0(\vec{k})}} < 0|\left(a(-\vec{p})+a^{\dagger}(\vec{p})\right)$$

$$\times e^{i(H_0-E_0)(t_2-t_1)}\left(a(-\vec{k})+a^{\dagger}(\vec{k})\right)|0 >$$

$$= \sum_{\vec{k}} \frac{\tilde{J}(-\vec{k},t_1)\tilde{J}(\vec{k},t_2)}{2k_0(\vec{k})}e^{-ik_0(\vec{k})(t_1-t_2)}. \quad (5.10)$$

Combining these results we find

$$< 0|\Psi(t) > e^{iE_0t}$$

$$= 1 - \bar{\varepsilon}^2 \sum_{\vec{k}} \int_0^t dt_1 \int_0^{t_1} dt_2 \frac{\tilde{J}(-\vec{k},t_1)\tilde{J}(\vec{k},t_2)}{2k_0(\vec{k})}e^{-ik_0(\vec{k})(t_1-t_2)} + \mathcal{O}(\bar{\varepsilon}^3)$$

$$= 1 - \tfrac{1}{2}\bar{\varepsilon}^2 \sum_{\vec{k}} \int_0^t dt_1 \int_0^t dt_2 \frac{\tilde{J}(-\vec{k},t_1)\tilde{J}(\vec{k},t_2)}{2k_0(\vec{k})}e^{-ik_0(\vec{k})|t_1-t_2|} + \mathcal{O}(\bar{\varepsilon}^3). \quad (5.11)$$

Especially the last identity is useful to relate this to the Green's function we introduced in the previous section. Using contour deformation in the complex ω plane we find

$$\int_{-\infty}^{\infty} d\omega \frac{e^{i\omega t}}{\omega^2 - k_0^2(\vec{k}) + i\varepsilon} = -\frac{2\pi i}{2k_0(\vec{k})}e^{-ik_0(\vec{k})|t|}. \quad (5.12)$$

This can be shown as follows. When $t > 0$, we can deform the contour of integration to the upper half-plane (where $e^{i\omega t}$ decays exponentially) and only the pole at $\omega = \omega_- \equiv -k_0(\vec{k})+i\varepsilon$ contributes, with a residue $2\pi i e^{-ik_0(\vec{k})t}/[-2k_0(\vec{k})]$ (see Figure 5.1). Instead, for $t < 0$ the contour needs to be deformed to the lower half-plane and the pole at $\omega = \omega_+ \equiv k_0(\vec{k}) - i\varepsilon$ contributes with the residue $2\pi i e^{ik_0(\vec{k})t}/[-2k_0(\vec{k})]$ (note that the contour now runs clockwise, giving an extra minus sign).

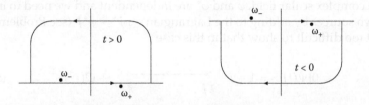

FIGURE 5.1
Contour deformation to define the integration.

This means that we can rewrite Equation (5.11) as

$$\lim_{t\to\infty} <0|\Psi(t)> e^{iE_0t} = 1 - \frac{i}{2}\bar{\varepsilon}^2 \sum_{\vec{k}} \int_{-\infty}^{\infty} d\omega \, \frac{\tilde{J}(-\omega,-\vec{k})\tilde{J}(\omega,\vec{k})}{\omega^2 - \vec{k}^2 - m^2 + i\varepsilon} + \mathcal{O}(\bar{\varepsilon}^3),$$

(5.13)

where

$$\tilde{J}(\omega,\vec{k}) = \frac{1}{\sqrt{2\pi}}\int dt \, \tilde{J}(\vec{k},t)e^{i\omega t} = \frac{1}{\sqrt{2\pi V}}\int_{\mathbb{R}\times V} d_3\vec{x}dt \, J(x)e^{ikx}. \quad (5.14)$$

The last expression should be replaced by $(2\pi)^{-2}\int_{\mathbb{R}^4} d_4x \, J(x)e^{ikx}$ in case the volume is infinite. It is important to note that we have chosen $J(x) = 0$ for $t < 0$. Equivalently we can start at $t = -\infty$ and integrate the quantum equation of motion up to $t = \infty$. We have to require that $J(x)$ vanishes sufficiently rapidly at infinity.

In an infinite volume we therefore find for what is known as the vacuum to vacuum amplitude of the scattering matrix

$$\lim_{t\to\infty} <0|\Psi(t)> e^{iE_0t} = 1 - \frac{i}{2}\bar{\varepsilon}^2 \int d_4k \, \frac{\tilde{J}(-k)\tilde{J}(k)}{k^2 - m^2 + i\varepsilon} + \mathcal{O}(\bar{\varepsilon}^3)$$

$$= 1 - \frac{i}{2}\bar{\varepsilon}^2 \int d_4xd_4y \, J(x)G(x-y)J(y) + \mathcal{O}(\bar{\varepsilon}^3).$$

(5.15)

where $G(x-y)$ is exactly the Green's function we introduced in the previous section. The so-called $i\varepsilon$ prescription, which is equivalent with specifying the boundary conditions, has therefore been derived from the time ordering in the Hamiltonian evolution of the system and is thus prescribed by the requirement of *causality*. Note that we can use the diagrams introduced in the previous section to express this result (taking $E_0 = 0$ from now on) as

$$<0|\Psi(t)> = 1 - \frac{i}{2} \quad \overset{\times}{\underset{\bar{\varepsilon}J}{\rule{3cm}{0.4pt}}}\overset{\times}{\underset{\bar{\varepsilon}J}{}} \quad + \mathcal{O}(\bar{\varepsilon}^3),$$

(5.16)

where the factor of a half is a consequence of the symmetry under interchanging the two sources.

For a complex scalar field, φ and φ^* are independent and we need to introduce two sources by adding to the Lagrangian $-\varphi J^* - \varphi^* J$ (see Problem 17). It is not too difficult to show that in this case

$$<0|\Psi(t)> = 1 - i \quad \overset{\times}{\underset{\bar{\varepsilon}J}{\rule{3cm}{0.4pt}}}\overset{\times}{\underset{\bar{\varepsilon}J^*}{}} \quad + \mathcal{O}(\bar{\varepsilon}^3),$$

(5.17)

without a factor of one half because the sources J and J^* are independent and cannot be interchanged. This is why in this case the propagator has a direction.

6

Path Integrals in Quantum Mechanics

DOI: 10.1201/b15364-6

For simplicity we will start with a one-dimensional Hamiltonian

$$H = \frac{\hat{p}^2}{2m} + V(\hat{x}), \quad \hat{p} = \frac{\hbar}{i}\frac{\partial}{\partial x}, \tag{6.1}$$

where we have indicated a hat on top of operators to distinguish them from number-valued coordinate and momentum. We wish to study the time-evolution operator $\exp(-iHT/\hbar)$. In the coordinate representation its matrix is given by

$$< x'|e^{-iHT/\hbar}|x >, \tag{6.2}$$

where $|x>$ is the position eigenfunction. We will also need the momentum eigenfunction $|p>$, i.e., $\hat{p}|p> = p|p>$, whose wave function in the coordinate space is given by

$$< x|p > = \frac{e^{ipx/\hbar}}{\sqrt{2\pi\hbar}}. \tag{6.3}$$

Indeed, one verifies that

$$\frac{\hbar}{i}\frac{\partial}{\partial x} < x|p > = p < x|p >. \tag{6.4}$$

An important ingredient in deriving the path integral expression will be the completeness relations

$$\hat{1} = \int dx\, |x><x| \quad \text{and} \quad \hat{1} = \int dp\, |p><p|. \tag{6.5}$$

For arbitrary N we can use this to write

$$< x'|e^{-iHT/\hbar}|x > = < x'|\left(e^{-iHT/N\hbar}\right)^N|x >$$

$$= \int \cdots \int < x'|e^{-iHT/N\hbar}|x_{N-1} > dx_{N-1}$$

$$< x_{N-1}|e^{-iHT/N\hbar}|x_{N-2} > dx_{N-2}$$

$$< x_{N-2}|\cdots\cdots e^{-iHT/N\hbar}|x_2 > dx_2$$

$$< x_2|e^{-iHT/N\hbar}|x_1 > dx_1 < x_1|e^{-iHT/N\hbar}|x >. \tag{6.6}$$

We will now use the so-called Trotter formula

$$e^{-i(A+B)/N} = e^{-iA/N}e^{-iB/N}(1 + \mathcal{O}(N^{-2})) \qquad (6.7)$$

for two operators A and B. This can be seen by expanding the exponents, and the error term is actually of the form $[A, B]/N^2$. (One can also use the Campbell–Baker–Hausdorff formula, which will be introduced later). With the Hamiltonian of Equation (6.1) this can be used to write for $N \to \infty$

$$e^{-iHT/N\hbar} = e^{-i\hat{p}^2T/2mN\hbar}e^{-iV(\hat{x})T/N\hbar}. \qquad (6.8)$$

By inserting the completeness relation for the momentum we can eliminate the operators

$$
\begin{aligned}
< x_{i+1}|e^{-iHT/N\hbar}|x_i > &= \int dp_i \; < x_{i+1}|p_i >< p_i|e^{-iHT/N\hbar}|x_i > \\
&\approx \int dp_i \; < x_{i+1}|p_i >< p_i|e^{-i\hat{p}^2T/2mN\hbar}e^{-iV(\hat{x})T/N\hbar}|x_i > \\
&= \int dp_i \; < x_{i+1}|p_i >< p_i|e^{-ip_i^2T/2mN\hbar}e^{-iV(x_i)T/N\hbar}|x_i > \\
&= \int dp_i \; \frac{e^{ip_i(x_{i+1}-x_i)/\hbar}}{2\pi\hbar}e^{-i\left[\frac{p_i^2}{2m}+V(x_i)\right]T/N\hbar}.
\end{aligned} \qquad (6.9)
$$

This can be done for each matrix element occurring in Equation (6.6). Writing $\Delta t = T/N$, $x_N = x'$ and $x_0 = x$ we find

$$
\begin{aligned}
< x'|e^{-iHT/\hbar}|x > &= \lim_{N\to\infty} \int \cdots \int \prod_{i=1}^{N-1} dx_i \prod_{j=0}^{N-1} dp_j \\
&\quad \times \prod_{i=0}^{N-1} < x_{i+1}|p_i >< p_i|e^{-iH\Delta t/\hbar}|x_i > \\
&= \lim_{N\to\infty} \int \frac{dp_0}{2\pi\hbar} \prod_{i=1}^{N-1} \int \frac{dx_i dp_i}{2\pi\hbar} \\
&\quad \times \exp\left[\frac{i\Delta t}{\hbar}\sum_{i=0}^{N-1}\left(\frac{p_i(x_{i+1}-x_i)}{\Delta t}-\frac{p_i^2}{2m}-V(x_i)\right)\right].
\end{aligned} \qquad (6.10)
$$

It is important to observe that there is one more p integration than the number of x integrations.

The integrals in the path integral are strongly oscillating and can only be defined by analytic continuation. As parameter for this analytic continuation, one chooses the time t. For $\Delta t = T/N \equiv -iT/N = -i\Delta\tau$, the Gaussian integral over the momenta is easily evaluated

$$\int_{-\infty}^{\infty} dp_i \, \exp\left[\frac{ip_i(x_{i+1} - x_i)}{\hbar} - \frac{p_i^2 \Delta\tau}{2m\hbar}\right]$$

$$= \sqrt{2\pi m\hbar/\Delta\tau} \, \exp\left[-\tfrac{1}{2}m\frac{(x_{i+1} - x_i)^2}{\Delta\tau\hbar}\right], \qquad (6.11)$$

which leads to

$$< x'|e^{-iHT/\hbar}|x > = \lim_{N\to\infty} \left(\frac{mN}{2\pi T\hbar}\right)^{\frac{1}{2}N}$$

$$\times \int \prod_{i=1}^{N-1} dx_i \, \exp\left[-\frac{\Delta\tau}{\hbar} \sum_{i=0}^{N-1} \left(\frac{m(x_{i+1} - x_i)^2}{2\Delta\tau^2} + V(x_i)\right)\right] \qquad (6.12)$$

or after substituting $T = iT$ we find

$$< x'|e^{-iHT/\hbar}|x > = \lim_{N\to\infty} \left(\frac{mN}{2\pi i T\hbar}\right)^{\frac{1}{2}N}$$

$$\times \int \prod_{i=1}^{N-1} dx_i \, \exp\left[i\frac{\Delta t}{\hbar} \sum_{i=0}^{N-1} \left(\frac{m(x_{i+1} - x_i)^2}{2\Delta t^2} - V(x_i)\right)\right]. \qquad (6.13)$$

This is the definition of the path integral, but formally it will often be written as

$$< x'|e^{-iHT/\hbar}|x > = \int_{x(0)=x}^{x(T)=x'} \mathcal{D}x(t) \, \exp[i\,S/\hbar],$$

$$S = \int_0^T dt \, \{\tfrac{1}{2}m\dot{x}^2(t) - V(x(t))\}, \qquad (6.14)$$

since the discretised version of the action with $x_j \equiv x(t = j\Delta t)$ is precisely

$$S_{\text{discrete}} = \Delta t \sum_{i=0}^{N-1} \left(\tfrac{1}{2}m\frac{(x_{i+1} - x_i)^2}{\Delta t^2} - V(x_i)\right). \qquad (6.15)$$

It is important to note that the continuous expression is just a notation for the discrete version of the path integral, but formal manipulations will be much easier to perform in this continuous formulation. Furthermore, the integral is only defined through the analytic continuation in time.

However, if we integrate over $x_N = x_0$ this analytically continued path integral, with $T = -iT$, has an important physical interpretation

$$\int dx \, < x|e^{-HT/\hbar}|x > = \text{Tr}(e^{-\beta H})_{\beta=T/\hbar}. \qquad (6.16)$$

It is the quantum thermal partition function (the Boltzmann distribution) with a temperature of $\hbar/kT$. In the continuous formulation we therefore have

$$\mathrm{Tr}(e^{-TH/\hbar}) = \int_{x(T)=x(0)} \mathcal{D}x(\tau) \, \exp[-S_E/\hbar] \tag{6.17}$$

in which S_E is the so-called Euclidean action

$$S_E = \int_0^T d\tau \left\{ \tfrac{1}{2}m(dx(\tau)/d\tau)^2 + V(x(\tau)) \right\}. \tag{6.18}$$

It is only in this Euclidean case that one can define the path integral in a mathematically rigorous fashion on the class of piecewise continuous functions in terms of the so-called Wiener measure

$$\int dW_{x,x'}(T) \equiv \int_{x(0)=x}^{x(T)=x'} \mathcal{D}x(\tau) \exp\left(-\int_0^T \tfrac{1}{2}m\dot{x}(\tau)^2 d\tau/\hbar \right), \tag{6.19}$$

meaning that this measure is independent of the way the path is discretised, Figure 6.1.

For more details on this, see *Quantum Physics: A Functional Integral Point of View*, by J. Glimm and A. Jaffe (2nd ed., Springer, New York, 1987).

We will now do an exact computation to give us some confidence in the formalism. To be specific, what we will compute is the quantum partition function for the harmonic oscillator, where $V(x) = \tfrac{1}{2}m\omega^2 x^2$

$$Z_N \equiv \left(\frac{mN}{2\pi T\hbar} \right)^{\frac{1}{2}N} \int_{-\infty}^{\infty} \cdots \int \prod_{i=0}^{N-1} dx_i$$

$$\times \exp\left[-\frac{\Delta\tau}{\hbar} \sum_{i=0}^{N-1} \tfrac{1}{2}m \left(\frac{x_{i+1}-x_i}{\Delta\tau} \right)^2 + \tfrac{1}{2}m\omega^2 x_i^2 \right]. \tag{6.20}$$

Note that we have now N integrations, because we also integrate over $x_0 = x(0) = x(T) = x_N$ to implement the trace. The path involved is thus periodic in time, a general feature of the expression for the quantum partition function

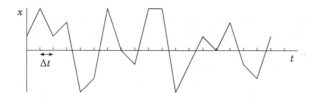

FIGURE 6.1
Contributing path.

in terms of a path integral. We now rescale

$$y_i = x_i \left(\frac{m}{\Delta \tau \hbar} \right)^{\frac{1}{2}}, \qquad \tilde{\omega} = \omega \Delta \tau, \tag{6.21}$$

to obtain the simple result

$$Z_N = \int_{-\infty}^{\infty} \cdots \int \prod_{i=0}^{N-1} \frac{dy_i}{\sqrt{2\pi}} \, \exp\left[-\sum_{i=0}^{N-1} \tfrac{1}{2}(y_{i+1} - y_i)^2 + \tfrac{1}{2}\tilde{\omega}^2 y_i^2 \right]. \tag{6.22}$$

We can diagonalise the quadratic term by using Fourier transformation

$$y_k = \frac{1}{\sqrt{N}} \sum_{\ell=0}^{N-1} b_\ell e^{2\pi i k \ell / N}, \qquad b_\ell^* = b_{N-\ell}, \qquad b_0^* = b_0. \tag{6.23}$$

It is easy to verify that the Jacobian for the change of variables $y_i \to b_\ell$ is 1, and one obtains a result that must look familiar from the classical small oscillations problem for a finite number of weights connected by strings,

$$Z_N = \int_{-\infty}^{\infty} \cdots \int \prod_{\ell=0}^{N-1} \frac{db_\ell}{\sqrt{2\pi}} \, \exp\left[-\tfrac{1}{2} \sum_{\ell=0}^{N-1} \left(4\sin^2(\pi \ell / N) + \tilde{\omega}^2 \right) |b_\ell|^2 \right]. \tag{6.24}$$

Note that if b_ℓ is complex, we mean by $db_\ell = d\mathrm{Re}b_\ell \, d\mathrm{Im}b_\ell$. The integral can now easily be evaluated

$$Z_N = \prod_{\ell=0}^{N-1} \left(4\sin^2(\pi \ell / N) + \tilde{\omega}^2 \right)^{-\frac{1}{2}}. \tag{6.25}$$

We can convert the product to a sum using a Laplace transform. We start with the identity

$$\log(\lambda/\mu) = -\lim_{a \searrow 0} \int_0^\infty ds \, s^{a-1}(e^{-s\lambda} - e^{-s\mu}), \tag{6.26}$$

such that

$$\log[Z_N(\tilde{\omega})/Z_N(\tilde{\omega}_0)] = \tfrac{1}{2} \lim_{a \searrow 0} \int_0^\infty ds \, s^{-1+a} \left(Q(s, \tilde{\omega}) - Q(s, \tilde{\omega}_0) \right). \tag{6.27}$$

We read off, from the definition of Z_N, that Q is a sum of exponentials

$$Q(s, \tilde{\omega}) \equiv \sum_{\ell=0}^{N-1} \exp\left[-s \left(4\sin^2(\pi \ell / N) + \tilde{\omega}^2 \right) \right] \equiv N e^{-s(\tilde{\omega}^2 + 2)} f_s(0), \tag{6.28}$$

where

$$f_s(x) = \frac{1}{N} \sum_{\ell=0}^{N-1} e^{2s \cos(2\pi(\ell+x)/N)}. \tag{6.29}$$

This is a periodic function with period 1 $[f_s(x+1) = f_s(x)]$, and its discrete Fourier coefficients can be computed exactly

$$\tilde{f}_s(k) = \int_0^1 dx \, e^{2\pi i k x} f_s(x) = \frac{1}{N} \int_0^N dx \, e^{2\pi i k x} e^{2s \cos(2\pi x/N)}$$

$$= \frac{1}{2\pi} \int_0^{2\pi} d\theta \, e^{iNk\theta} e^{2s \cos(\theta)} \equiv I_{Nk}(2s). \tag{6.30}$$

Please note that we have exchanged the sum over ℓ with extending the integration of x to the interval $[0, N]$. The last identity is one of the definitions of the modified Bessel function, see, e.g., *Handbook of Mathematical Functions*, by M. Abramowitz and I. Stegun (Dover, New York, 1978). The advantage of these manipulations is that the Laplace transform of this Bessel function is now (see the same reference)

$$\int_0^\infty ds \, e^{-\lambda s} I_\nu(s) = \frac{[\lambda + \sqrt{\lambda^2 - 1}]^{-|\nu|}}{\sqrt{\lambda^2 - 1}}, \tag{6.31}$$

and as we can express $f_s(x)$ as a sum over these Bessel functions

$$f_s(x) = \sum_{k \in \mathbb{Z}} e^{-2\pi i k x} I_{Nk}(2s), \tag{6.32}$$

this allows us to evaluate Equation (6.27). For technical reasons, it is easier to compute the variation of the free energy with the frequency, where the free energy F is defined as

$$Z_N(\tilde{\omega}) = \exp[-\beta F(\tilde{\omega})]. \tag{6.33}$$

We therefore find (using $\beta = T/\hbar = N\Delta\tau/\hbar$)

$$\frac{\partial}{\partial \tilde{\omega}} F(\tilde{\omega}) = \hbar\omega \sum_{k \in \mathbb{Z}} \int_0^\infty ds \, e^{-(2+\tilde{\omega}^2)s} I_{Nk}(2s)$$

$$= \frac{\hbar\omega}{2\sqrt{(\frac{1}{2}\tilde{\omega}^2 + 1)^2 - 1}} \sum_{k \in \mathbb{Z}} \left[\frac{1}{2}\tilde{\omega}^2 + 1 + \sqrt{(\frac{1}{2}\tilde{\omega}^2 + 1)^2 - 1} \right]^{-N|k|}. \tag{6.34}$$

The geometric series is of course easily summed, but to make the result more transparent we introduce the scaled *effective* frequency Ω

$$\omega T/N = \omega\Delta\tau = \tilde{\omega} \equiv 2\sinh(\tfrac{1}{2}\Omega) \tag{6.35}$$

and using the identity $\frac{1}{2}\tilde{\omega}^2 + 1 = \cosh(\Omega)$, we easily find that

$$\frac{\Delta\tau}{\hbar} \frac{\partial}{\partial\Omega} F(\Omega) = \frac{1}{2} \sum_{k \in \mathbb{Z}} e^{-N\Omega|k|} = \frac{1}{1 - e^{-N\Omega}} - \frac{1}{2}$$

$$= -\frac{1}{N} \frac{\partial}{\partial\Omega} \log\left(\frac{e^{-\Omega N/2}}{1 - e^{-N\Omega}} \right). \tag{6.36}$$

The last identity can be seen as the free energy of the harmonic oscillator with the frequency $\Omega N/T = \Omega/\Delta\tau$, as it can also be written as

$$\frac{\Delta\tau}{\hbar}\frac{\partial}{\partial\Omega}F(\Omega) = -\frac{1}{N}\frac{\partial}{\partial\Omega}\log\left(\sum_{n=0}^{\infty}e^{-(n+\frac{1}{2})\Omega N}\right). \tag{6.37}$$

Amazingly, even at finite N the Euclidean path integral agrees with the quantum partition function of a harmonic oscillator, but with a frequency that is modified by the discretisation; see Eqs. (6.21) and (6.35). It is trivial to check now that the limit $N \to \infty$ is well defined and gives the required result, since

$$\lim_{N\to\infty}\Omega N/T = \omega. \tag{6.38}$$

In general the exact finite N path integral is no longer of a simple form. Nevertheless, one can evaluate this exact expression in relatively simple terms (which will verify the above result along a different route; see also Problem 10). So from now on, we will take the potential arbitrary and in a sense we follow the derivation of the path integral in the reverse order.

$$Z_N(x', x; T) = \int\frac{dp_0}{2\pi\hbar}\prod_{i=1}^{N-1}\int\frac{dx_i dp_i}{2\pi\hbar}$$

$$\times \exp\left[\frac{\Delta\tau}{\hbar}\sum_{i=0}^{N-1}\left(\frac{ip_i(x_{i+1}-x_i)}{\Delta\tau}-\frac{p_i^2}{2m}-V(x_i)\right)\right]$$

$$= \int dp_0\prod_{i=1}^{N-1}\int dx_i dp_i\prod_{j=0}^{N-1}<x_{j+1}|p_j><p_j|\exp\left(-\frac{\Delta\tau\hat{p}^2}{2m\hbar}\right)$$

$$\times \exp\left(-\frac{\Delta\tau V(\hat{x})}{\hbar}\right)|x_j>$$

$$= <x'|\left\{\exp\left(-\frac{\Delta\tau\hat{p}^2}{2m\hbar}\right)\exp\left(-\frac{\Delta\tau V(\hat{x})}{\hbar}\right)\right\}^N|x>. \tag{6.39}$$

This means that we can define an *effective* Hamiltonian by

$$e^{-H(N)T/\hbar} \equiv \left\{\exp\left(-\frac{\Delta\tau\hat{p}^2}{2m\hbar}\right)\exp\left(-\frac{\Delta\tau V(\hat{x})}{\hbar}\right)\right\}^N. \tag{6.40}$$

But this Hamiltonian is not Hermitian as one easily checks from the above expression, since under conjugation the order of the exponents containing the kinetic and potential terms is reversed. This can be corrected in two ways

$$e^{-H_1(N)\Delta\tau/\hbar} \equiv \exp\left(-\frac{\Delta\tau\hat{p}^2}{4m\hbar}\right)\exp\left(-\frac{\Delta\tau V(\hat{x})}{\hbar}\right)\exp\left(-\frac{\Delta\tau\hat{p}^2}{4m\hbar}\right),$$

$$e^{-H_2(N)\Delta\tau/\hbar} \equiv \exp\left(-\frac{\Delta\tau V(\hat{x})}{2\hbar}\right)\exp\left(-\frac{\Delta\tau\hat{p}^2}{2m\hbar}\right)\exp\left(-\frac{\Delta\tau V(\hat{x})}{2\hbar}\right), \tag{6.41}$$

leading to two equivalent expressions for the finite N path integral

$$Z_N(x', x; T) = < x' | \exp\left(-\frac{\Delta\tau \hat{p}^2}{4m\hbar}\right) \exp\left(-H_1(N)T/\hbar\right) \exp\left(\frac{\Delta\tau \hat{p}^2}{4m\hbar}\right) |x >$$

$$Z_N(x', x; T) = < x' | \exp\left(\frac{\Delta\tau V(\hat{x})}{2\hbar}\right) \exp\left(-H_2(N)T/\hbar\right) \exp\left(-\frac{\Delta\tau V(\hat{x})}{2\hbar}\right) |x >.$$

$$(6.42)$$

In particular the partition function is given by

$$Z_N = \int dx \, Z_N(x, x, T) = \text{Tr}(e^{-H_1(N)T/\hbar}) = \text{Tr}(e^{-H_2(N)T/\hbar}). \qquad (6.43)$$

It is actually not too difficult to show that there exists a unitary transformation U, such that $U H_1 U^\dagger = H_2$, which shows that both choices are indeed physically equivalent.

In principle we can now compute $H_i(N)$ for finite N as an expansion in $1/N$, by using the so-called Campbell–Baker–Hausdorff formula

$$e^A e^B = e^{F(A,B)}, \quad F(A, B) = A + B + \tfrac{1}{2}[A, B] + \frac{1}{12}[A, [A, B]]$$

$$+ \frac{1}{12}[B, [B, A]] + \cdots, \qquad (6.44)$$

which is a series in multiple commutators of the, in general, noncommuting operators A and B. It can be derived by expanding the exponentials, but in the mathematics literature more elegant constructions are known, based on properties of Lie groups and Lie algebras. These objects will be discussed in Chapter 18. For the harmonic oscillator, working out the products of the exponential can be done to all orders and one finds (see Problem 10 for details)

$$\Delta\tau H_i(N) = \frac{\hat{p}^2}{2M_i} + \tfrac{1}{2}M_i \Omega^2 \hat{x}^2, \qquad (6.45)$$

with Ω defined as in Equation (6.35) and the *effective* masses M_i defined by

$$M_1 = \frac{2m \tanh(\tfrac{1}{2}\Omega)}{\Delta\tau \Omega}, \quad M_2 = \frac{m \sinh(\Omega)}{\Delta\tau \Omega}. \qquad (6.46)$$

One can now explicitly verify that [compare this to Equation (6.37)]

$$\text{Tr}\big(\exp(-H_1(N)T/\hbar)\big) = \text{Tr}\big(\exp(-H_2(N)T/\hbar)\big) = \exp\big(-F(\Omega)T/\hbar\big). \quad (6.47)$$

Now we have seen that, at least for some examples, the limit of increasingly finer discretisation is in principle well defined, we can think of generalisation to an arbitrary number of dimensions (n) (for field theory even to an infinite number of dimensions).

$$< \vec{x}' | e^{-iHT/\hbar} | \vec{x} > \equiv \int \mathcal{D}\vec{x}(t) \, \exp \left[\frac{i}{\hbar} \int_0^T dt \, L(\vec{x}(t), \dot{\vec{x}}(t)) \right]$$

$$= \lim_{N \to \infty} \int \frac{d_n p_0}{(2\pi\hbar)^n} \int \prod_{i=1}^{N-1} \frac{d_n x_i d_n p_i}{(2\pi\hbar)^n}$$

$$\times \exp \left[i \frac{\Delta t}{\hbar} \sum_{i=0}^{N-1} \left(\vec{p}_i \cdot \frac{(\vec{x}_{i+1} - \vec{x}_i)}{\Delta t} - H(\vec{p}_i, \vec{x}_i) \right) \right]$$

$$= \lim_{N \to \infty} \left(\frac{mN}{2\pi i T\hbar} \right)^{\frac{1}{2}nN}$$

$$\times \int \prod_{i=1}^{N-1} d_n x_i \, \exp \left[i \frac{\Delta t}{\hbar} \sum_{i=0}^{N-1} L(\vec{x}_i, (\vec{x}_{i+1} - \vec{x}_i)/\Delta t) \right]. \quad (6.48)$$

We have purposely also given the expression that involves the path integral as an integral over phase space, as it shows that the Gaussian integration over the momenta effectuates the Legendre transform

$$i\vec{p} \cdot \dot{\vec{x}} - i\frac{\vec{p}^2}{2m} - iV(\vec{x}) = -i\frac{(\vec{p} - m\dot{\vec{x}})^2}{2m} + i\frac{m\dot{\vec{x}}^2}{2} - iV(\vec{x}), \quad (6.49)$$

which is equivalent to the stationary phase approximation for the momentum integration

$$\frac{\delta}{\delta\vec{p}} (\vec{p} \cdot \dot{\vec{x}} - H(\vec{p}, \vec{x})) = \dot{\vec{x}} - \frac{\partial H}{\partial \vec{p}} = \dot{\vec{x}} - \frac{\vec{p}}{m} = 0. \quad (6.50)$$

An other interesting example of the path integral is the case of the interaction of a charged particle with a magnetic field. In that case one has for the Hamiltonian

$$H(\hat{\vec{p}}, \hat{\vec{x}}) = \frac{(\hat{\vec{p}} - e\vec{A}(\hat{\vec{x}}))^2}{2m} + V(\hat{\vec{x}}). \quad (6.51)$$

Now, however, the matrix element $< \vec{p}_i | \exp(-iH\Delta t/\hbar) | \vec{x}_i >$ will depend on the specific ordering for the position and momentum operators in H. Different orderings differ by terms linear in $\hbar$, or

$$\vec{A}(\hat{\vec{x}}) \cdot \hat{\vec{p}} = \hat{\vec{p}} \cdot \vec{A}(\hat{\vec{x}}) + i\hbar\partial_i A_i(\hat{\vec{x}}). \quad (6.52)$$

So, by choosing the so-called Coulomb gauge $\partial_i A_i(\vec{x}) = 0$, the problem of operator ordering disappears. We leave it as an exercise to verify that the action, obtained from the Legendre transform, is given by

$$S = \int_0^T dt \left(\tfrac{1}{2} m \dot{\vec{x}}^2 - V(\vec{x}) + e \dot{\vec{x}} \cdot \vec{A}(\vec{x}) \right). \tag{6.53}$$

Under a gauge transformation $A_i(\vec{x}) \rightarrow A_i(\vec{x}) + \partial_i \Lambda(\vec{x})$, one finds that the action changes to $S_\Lambda = S + e\{\Lambda[\vec{x}(T)] - \Lambda[\vec{x}(0)]\}$. Using the path integral this means that

$$< \vec{x}' | \exp(-i H_\Lambda T/\hbar) | \vec{x} > \, = \exp\left(i e \Lambda(\vec{x}')/\hbar \right) < \vec{x}' |$$
$$\times \exp(-i H T/\hbar) | \vec{x} > \exp\left(- i e \Lambda(\vec{x})/\hbar \right). \tag{6.54}$$

Since it is easily shown that $H_\Lambda = \exp\left[i e \Lambda(\hat{\vec{x}})/\hbar \right] H \exp\left[- i e \Lambda(\hat{\vec{x}})/\hbar \right]$, this proves that the path integral derived from Equation (6.53) has the correct properties under gauge transformations, despite the fact that the derivation was performed by first going to the Coulomb gauge.

As long as the Hamiltonian is quadratic in the momenta, the stationary phase approximation for the momentum integral is exact. However, also for the coordinate integrals we can use the stationary phase *approximation* (exact for a harmonic oscillator), which is related to the WKB approximation in quantum mechanics. It gives a way of defining an expansion in $\hbar$, where in accordance to the correspondence principle, the lowest-order term reproduces the classical time evolution. Indeed the stationary phase condition

$$\frac{\delta S}{\delta x^i(t)} = \frac{\delta S}{\delta x^i(t)} - \frac{d}{dt} \frac{\delta S}{\delta \dot{x}^i(t)} = 0 \tag{6.55}$$

is precisely solved by the classical solutions, $\vec{x}_{cl}(t)$, with $\vec{x}_{cl}(0) = \vec{x}$ and $\vec{x}_{cl}(T) = \vec{x}'$. We expand around these solutions by writing

$$\vec{x}(t) = \vec{x}_{cl}(t) + \vec{q}(t), \quad \vec{q}(0) = \vec{q}(T) = \vec{0}, \tag{6.56}$$

such that

$$S(\vec{x}) = S(\vec{x}_{cl}) + \tfrac{1}{2} \int dt' dt \, q^i(t) \frac{\delta^2 S(\vec{x}_{cl})}{\delta q^i(t) \delta q^j(t')} q^j(t') + \mathcal{O}(q^3). \tag{6.57}$$

There is no term linear in $q^i(t)$, as this term is proportional to the equations of motion, or equivalently to the stationary phase condition. For the simple Lagrangian $L = \tfrac{1}{2} m \dot{\vec{x}}^2 - V(\vec{x})$ one has

$$\frac{\delta^2 S(\vec{x}_{cl})}{\delta q^i(t) \delta q^j(t')} = -\delta(t - t') \left(m \delta_{ij} \frac{d^2}{dt^2} + M_{ij}(t) \right),$$
$$M_{ij}(t) \equiv \frac{\partial^2 V(\vec{x})}{\partial x^i \partial x^j} \Big|_{\vec{x} = \vec{x}_{cl}(t)}. \tag{6.58}$$

For the harmonic potential, $V = \frac{1}{2}m\omega^2\vec{x}^2$, where the stationary phase approximation is exact, i.e., there are no $\mathcal{O}(q^3)$ corrections. Introducing $\vec{q}(t)$, however, splits the action in a classical piece that depends on the boundary conditions and a quantum piece described by a harmonic oscillator action for the fluctuations around the classical path that is independent of the boundary conditions *and* the classical path

$$S(\vec{x}) = S(\vec{x}_{cl}) + \int_0^T dt\, (\tfrac{1}{2}m\dot{\vec{q}}^{\,2} - \tfrac{1}{2}m\omega^2\vec{q}^{\,2}). \tag{6.59}$$

In practical situations one splits from the action the quadratic part in the coordinates and velocities and considers the rest as a perturbation. In that case x_{cl} is the classical solution of the quadratic part only. As this can always be solved exactly, and as nonquadratic path integrals can rarely be computed explicitly, this will be the way in which we will derive the Feynman rules for the quantum theory, in terms of which one can efficiently perform the perturbative computations.

7

Path Integrals in Field Theory

DOI: 10.1201/b15364-7

For a scalar field in a finite volume $V = [0, L]^3$, the Hamiltonian is given in the Fourier representation by [see Equation (2.6)]

$$H = \sum_{\vec{k}=2\pi\vec{n}/L} \left(\tfrac{1}{2}|\tilde{\pi}(\vec{k})|^2 + \tfrac{1}{2}(\vec{k}^2 + m^2)|\tilde{\varphi}(\vec{k})|^2 + V(\tilde{\varphi}) + \tilde{\varphi}(\vec{k})\tilde{J}(-\vec{k}, t) \right). \quad (7.1)$$

As $\varphi(x)$ is real we have $\tilde{\varphi}^*(\vec{k}) = \tilde{\varphi}(-\vec{k})$. It is customary to write the quadratic term in the fields (the mass term) explicitly, such that the potential $V(\varphi)$ only contains the interaction terms. If we like, we could split the Fourier modes in their real and imaginary components [the $\cos(\vec{x} \cdot \vec{k})$ and $\sin(\vec{x} \cdot \vec{k})$ modes]. Or even simpler is to use Dirichlet boundary conditions, i.e., $\varphi(x) = 0$ at the boundaries of the volume, such that the Fourier modes are given by $\prod_j \sin(\pi n_j x_j / L)$ (with $n_j > 0$), with real coefficients. In either case, for $V(\varphi) = 0$ the Hamiltonian simply describes an infinite set of decoupled harmonic oscillators, which can be truncated to a finite set by introducing a so-called momentum cutoff $|\vec{k}| \leq \Lambda$. In this case we know how to write the path integral, even in the presence of interactions. The introduction of a cutoff is called a regularisation. The field theory is called renormalisable if the limit $\Lambda \to \infty$ can be defined in a suitable way, often by varying the parameters in a suitable way with the cutoff. The class of renormalisable field theory is relatively small. For a finite momentum cutoff, the path integral is nothing but a simple generalisation of the one we defined for quantum mechanics in n dimensions, or in the absence of interactions

$$Z = \lim_{N \to \infty} \prod_{\vec{k}} (2\pi i \Delta t)^{-N/2} \int \prod_{j=1}^{N-1} \prod_{\vec{k}} d\tilde{\varphi}_j(\vec{k}) \exp\left[i\Delta t \sum_{j=0}^{N-1} \sum_{\vec{k}} \frac{|\tilde{\varphi}_{j+1}(\vec{k}) - \tilde{\varphi}_j(\vec{k})|^2}{2\Delta t^2} \right.$$

$$\left. - \tfrac{1}{2}(\vec{k}^2 + m^2)|\tilde{\varphi}_j(\vec{k})|^2 - \tilde{\varphi}_j(\vec{k})\tilde{J}(-\vec{k}, j\Delta t) \right]$$

$$\equiv \int \mathcal{D}\tilde{\varphi}(\vec{k}, t) \exp\left(i \int_0^T \sum_{\vec{k}} \{ \tfrac{1}{2}|\dot{\tilde{\varphi}}(\vec{k}, t)|^2 - \tfrac{1}{2}(\vec{k}^2 + m^2)|\tilde{\varphi}(\vec{k}, t)|^2 \right.$$

$$\left. - \tilde{\varphi}(\vec{k}, t)\tilde{J}(-\vec{k}, t) \} dt \right). \quad (7.2)$$

One of course identifies $\tilde{\varphi}_j(\vec{k}) = \tilde{\varphi}(\vec{k}, t = j\Delta t)$ and performing the Fourier transformation once more, one can write

$$\int_0^T dt \sum_{\vec{k}} \{\tfrac{1}{2}|\dot{\tilde{\varphi}}(\vec{k}, t)|^2 - \tfrac{1}{2}(\vec{k}^2 + m^2)|\tilde{\varphi}(\vec{k}, t)|^2 - \tilde{\varphi}(\vec{k}, t)\tilde{J}(-\vec{k}, t)\}$$

$$= \int_0^T dt \int_V d_3\vec{x}\{\tfrac{1}{2}(\partial_t\varphi(\vec{x}, t))^2 - \tfrac{1}{2}(\partial_i\varphi(\vec{x}, t))^2 - \tfrac{1}{2}m^2\varphi^2(\vec{x}, t) - \varphi(\vec{x}, t)J(\vec{x}, t)\}$$

$$= \int_{V\times[0,T]} d_4x \{\tfrac{1}{2}\partial_\mu\varphi(x)\partial^\mu\varphi(x) - \tfrac{1}{2}m^2\varphi^2(x) - \varphi(x)J(x)\}. \tag{7.3}$$

The last expression is manifestly Lorentz invariant apart from the dependence on the boundary conditions on the fields (which should disappear once we take L and T to infinity). This will allow us to perform perturbation theory in a Lorentz covariant way. (Things are somewhat subtle as any finite choice of the momentum cutoff does break the Lorentz invariance, and there are some theories where this is not restored when removing the cutoff, i.e., taking the limit $\Lambda \to \infty$.) This achieves a substantial simplification over Hamiltonian perturbation theory. It is now also trivial to reintroduce the interactions by adding the potential term to the Lagrange density, and we find in yet another shorthand notation for the measure the following expression for the path integral (implicitly assuming that the boundary values $\varphi(\vec{x}, 0)$ and $\varphi(\vec{x}, T)$ are fixed, prescribed functions)

$$Z = \int \mathcal{D}\varphi(x) \exp\left(i \int_{V\times[0,T]} d_4x\{\tfrac{1}{2}\partial_\mu\varphi(x)\partial^\mu\varphi(x) - \tfrac{1}{2}m^2\varphi^2(x)\right.$$
$$\left. - V(\varphi) - \varphi(x)J(x)\}\right). \tag{7.4}$$

In principle a path integral should be independent of the discretisation used in order to define it. For the Euclidean path integral, one particular way that is used quite often is the lattice discretisation, where instead of a momentum cutoff one makes not only time but also space discrete. This means that the field now lives on a lattice and its argument takes the values ja where $j \in \mathbb{Z}^4$ and a is the so-called lattice spacing, which in the end should be taken to zero. By suitably restricting the components of j, with appropriate boundary conditions on the fields, one keeps space and time finite, $V = a^3M^3$ and $T = aN$. This leads to an integral of the form

$$(2\pi a)^{-NM^3/2} \int \prod_j d\varphi_j \exp\left(-a^4 \sum_j \{\tfrac{1}{2}\sum_\mu \frac{(\varphi_{j+e_\mu} - \varphi_j)^2}{a^2}\right.$$
$$\left. + \tfrac{1}{2}m^2\varphi_j^2 + V(\varphi_j) + J_j\varphi_j\}\right) \tag{7.5}$$

where e_μ is a unit vector in the μ direction, and φ_j is identified with $\varphi(aj)$. In a sense, the momentum cutoff Λ is similar to the space-time cutoff $1/a$. The lattice formulation is very suitable for numerically evaluating the path integral, whereas the momentum cutoff is suitable for performing perturbation theory around the quadratic approximation of the action. For the latter we will compute, using the path integral, the same quantity as was calculated in Chapter 5, using Hamiltonian perturbation theory.

In the presence of a source, the Hamiltonian depends on time and the evolution operator has to be written in a way that takes the time ordering into account. As the time evolution operator $U(t)$ satisfies the Schrödinger equation

$$i\frac{d}{dt}U(t) = H(t)U(t), \quad U(0) = 1, \tag{7.6}$$

its solution can be written as (note the absence of $1/n!$)

$$U(t) = T\exp\left(-i\int_0^t H(t)dt\right)$$

$$\equiv \sum_{n=0}^{\infty}(-i)^n \int_0^t dt_1 \int_0^{t_1} \cdots \int_0^{t_{n-1}} dt_n\, H(t_1)\cdots H(t_n). \tag{7.7}$$

For convenience we introduce the notation

$$U(t_2, t_1) \equiv T\exp\left(-i\int_{t_1}^{t_2} H(t)dt\right), \quad t_2 > t_1 \tag{7.8}$$

which satisfies the property that

$$U(t_3, t_2)U(t_2, t_1) = U(t_3, t_1), \quad t_3 > t_2 > t_1. \tag{7.9}$$

In Chapter 5 we calculated the matrix element $< 0|U(T)|0 >$ to second order in the source (from now on we put $\bar\varepsilon = 1$). The Lagrangian relevant for the path integral evaluation is given by $\mathcal{L} = \frac{1}{2}\partial_\mu\varphi\partial^\mu\varphi - \frac{1}{2}m^2\varphi^2 - J\varphi$, with $J(x) = 0$ for $t < 0$ and $t > T$ (and for $\vec{x} \notin [0, L]^3$). The vacuum $|0 >$ is the state where all $\vec{k}$ oscillators are in their ground state. It turns out that we do not need an explicit expression for this vacuum wave functional, denoted by $\Psi_0(\{\tilde\varphi(\vec k)\}) \; = \; < \{\tilde\varphi(\vec k)\}|0>$. We have

$$
<0|U(T)|0> \; = \; \int \prod_{\vec p} d\tilde\varphi(\vec p)d\tilde\varphi'(\vec p) \; <0|\{\tilde\varphi'(\vec p)\}> \; < \{\tilde\varphi'(\vec p)\}|U(T)|\{\tilde\varphi(\vec p)\}>
$$
$$
\times < \{\tilde\varphi(\vec p)\}|0>, \tag{7.10}
$$

where $\{\tilde\varphi(\vec p)\}$ plays the role of x and $\{\tilde\varphi'(\vec p)\}$ the role of x'. The path integral expression for the evolution operator therefore becomes

$$
< \{\tilde\varphi'(\vec p)\}|U(T)|\{\tilde\varphi(\vec p)\}> \; = \; \int \mathcal{D}\varphi(x) \; \exp\left(i \int_0^T \int_V \mathcal{L}(\varphi(x))d_4x\right)
$$
$$
= \int \mathcal{D}\tilde\varphi(k) \; \exp\left(i \sum_{\vec k, k_0} \{\tfrac12(k^2 - m^2)|\tilde\varphi(k)|^2 \right.
$$
$$
\left. - \tilde\varphi(k)\tilde J(-k)\}\right). \tag{7.11}
$$

We have here also performed the Fourier transformation with respect to time, thereby converting the path integral measure $\mathcal{D}\tilde\varphi(\vec k, t)$ to the multiple integral over the (discrete temporal) Fourier components $\mathcal{D}\tilde\varphi(k)$, exactly as was done in one dimension [see Equation(6.23)], hence we also find a unit Jacobian for this change of variables. Obviously our notations are such that $k \equiv (k_0, \vec k)$ and $k^2 = k_0^2 - \vec k^2$. If we take the limit of space and time to infinity ($L \to \infty$ and $T \to \infty$), the sums over k can be converted into integrals. Finally we note that the oscillatory integrals occurring in the path integral can be dampened by replacing m^2 by $m^2 - i\varepsilon$ as this leads to replacement $\exp\left[i \int \mathcal{L}(\varphi)d_4x\right] \to \exp\left[i \int \mathcal{L}(\varphi)d_4x - \varepsilon \int \varphi^2 d_4x\right]$. This prescription also allows us to make the analytic continuation to imaginary time and coincides with the prescription derived for the propagator in the Hamiltonian formulation, so that causality is also properly implemented in the path integral approach. To recover the result obtained in the Hamiltonian approach, we simply split off a square

$$
< \{\tilde\varphi'(\vec p)\}|U(T)|\{\tilde\varphi(\vec p)\}>
$$
$$
= \int \mathcal{D}\tilde\varphi(k) \; \exp\left(i \sum_{\vec k, k_0} \tfrac12(k^2 - m^2 + i\varepsilon)\left|\tilde\varphi(k) - \frac{\tilde J(k)}{k^2 - m^2 + i\varepsilon}\right|^2\right)
$$
$$
\times \exp\left(-\frac{i}{2} \sum_k \frac{|\tilde J(k)|^2}{k^2 - m^2 + i\varepsilon}\right). \tag{7.12}
$$

We can now shift the integration of $\tilde{\varphi}(k)$ over $\tilde{J}(k)/(k^2 - m^2 + i\varepsilon)$ and introduce the Green's function in coordinate space [Equation (4.7)] to get

$$< \{\tilde{\varphi}'(\vec{p})\}|U(T)|\{\tilde{\varphi}(\vec{p})\} >$$

$$= \exp\left(-\frac{i}{2}\int d_4x\, d_4y\, J(x)G(x-y)J(y)\right)$$

$$\times \int D\tilde{\varphi}(k)\exp\left(i\sum_{\vec{k},k_0} \tfrac{1}{2}(k^2 - m^2 + i\varepsilon)|\tilde{\varphi}(k)|^2\right)$$

$$= \exp\left(-\frac{i}{2}\int d_4x\, d_4y\, J(x)G(x-y)J(y)\right) < \{\tilde{\varphi}'(\vec{p})\}|e^{-iH(J=0)T}|\{\tilde{\varphi}(\vec{p})\} > .$$

$$(7.13)$$

In the last step it is crucial to note that the source is taken to vanish for $t \leq 0$ and for $t \geq T$, as otherwise the shift we performed in the field would have changed the boundary values. Using Equation (7.10) we now obtain the remarkable result that

$$< 0|U(T)|0 > = < 0|e^{-iH(J=0)T}|0 > \exp\left(-\frac{i}{2}\int d_4x\, d_4y\, J(x)G(x-y)J(y)\right)$$

$$(7.14)$$

to *all* orders in the source J. Even at the level of a noninteracting scalar field theory, this demonstrates the dramatic simplifications that can arise from using the path integral method for calculating quantum amplitudes. One particular feature that is noteworthy in the path integral calculation is that the part of the Lagrangian that is quadratic in the fields represents the inverse propagator. This is no accident and is in general the way the (lowest-order) propagator is directly read off from the Lagrangian, since the quadratic part of the action is the starting point of the perturbative expansion. But before we will derive the Feynman rules from the perturbative expansion, it will be useful to emphasise that the time ordering, playing such an important role in the Hamiltonian formulation, is automatically implemented by the path integral. Furthermore, it will be helpful to understand in more detail how the source can be used to create and annihilate particles, as this will be our tool to write down the matrix elements of the evolution operator (the so-called scattering matrix, or for short, S-matrix) with respect to the basis specified by particle number and momentum (the so-called Fock space); see Equation (2.9).

In the Hamiltonian formulation we consider

$$< 0|\hat{\varphi}(\vec{x}', t_2)\hat{\varphi}(\vec{x}, t_1)|0 >, \qquad t_2 > t_1, \qquad (7.15)$$

where the field operator [compare this to Equation (2.8)] is given by

$$\hat{\varphi}(\vec{x}, t) = e^{iHt}\hat{\varphi}(\vec{x})e^{-iHt}, \qquad (7.16)$$

such that

$$< 0|\hat{\varphi}(\vec{x}', t_2)\hat{\varphi}(\vec{x}, t_1)|0 > = \frac{< 0|e^{-iH(T-t_2)}\hat{\varphi}(\vec{x}')e^{-iH(t_2-t_1)}\hat{\varphi}(\vec{x})e^{-iHt_1}|0 >}{< 0|e^{-iHT}|0 >},$$

(7.17)

where we have made use of the fact that the vacuum $|0 >$ is assumed to be an eigenstate of the Hamiltonian H (without a source term). The normalisation by $< 0|e^{-iHT}|0 >$ is hence a rather trivial factor. We can even write a similar expression in the presence of the source. In perturbation theory this would not be needed, but it is useful from a general point of view to consider this situation too.

In the presence of a time-dependent source one can write

$$< 0|U(T)\hat{\varphi}(\vec{x}', t_2)\hat{\varphi}(\vec{x}, t_1)|0 > = < 0|U(T, t_2)\hat{\varphi}(\vec{x}')U(t_2, t_1)\hat{\varphi}(\vec{x})U(t_1)|0 >.$$

(7.18)

In this case the field operators are of course given by

$$\hat{\varphi}(\vec{x}, t) = U(t)^\dagger \hat{\varphi}(\vec{x})U(t),$$

(7.19)

where in general $U(t)$ depends on the source J. It is now trivial, but a bit tedious, to convert this matrix element to a path integral. One first writes the product of the operators as a product of matrices in a suitable representation [e.g., the field representation $|\{\tilde{\varphi}(\vec{p})\} >$]. Each of the matrices for the three evolution operators can be written as a path integral, excluding the integral over the initial and final field components. The matrix product involves an integral over the final field component of the matrix to the right, which is also the initial field component for the matrix to the left. Without the insertion of the field operators $\hat{\varphi}(x)$, this would describe the fact that $U(t_3, t_2)U(t_2, t_1) = U(t_3, t_1)$ in the path integral formulation, which simply means that one *glues* the paths in $U(t_3, t_2)$ and $U(t_2, t_1)$ together by integrating over $\tilde{\varphi}(\vec{k}, t_2)$. With the field operator sandwiched between the two U matrices one simply includes its eigenvalue in the integrand over the paths, since the field operator (or its Fourier components) is diagonal on the states $|\{\tilde{\varphi}(\vec{p})\} >$. The final result can be written as

$$\int \prod_{\vec{p}} d\tilde{\varphi}(\vec{p})d\tilde{\varphi}'(\vec{p}) < 0|\{\tilde{\varphi}'(\vec{p})\} >$$

$$\times \left\{ \int \mathcal{D}\varphi(x) \exp(i \int_{t_2}^{T} \mathcal{L}\, d_4x)\varphi(\vec{x}', t_2) \exp(i \int_{t_1}^{t_2} \mathcal{L}\, d_4x) \right.$$

$$\left. \times \varphi(\vec{x}, t_1) \exp(i \int_0^{t_1} \mathcal{L}\, d_4x) \right\} < \{\tilde{\varphi}(\vec{p})\}|0 >,$$

(7.20)

where we implicitly assumed that the boundary conditions for the field φ in the path integral is in momentum space given by $\tilde{\varphi}(\vec{p}, t = 0) = \tilde{\varphi}(\vec{p})$ and $\tilde{\varphi}(\vec{p}, t = T) = \tilde{\varphi}'(\vec{p})$. The way the time ordering in the path integral is manifest is now obvious. Note that the field expectation values can also be written in terms of derivatives with respect to the sources, which is particularly simple to derive in the path integral formulation

$$< 0|U(T)\hat{\varphi}(\vec{x}', t_2)\hat{\varphi}(\vec{x}, t_1)|0 >= -\frac{\delta^2}{\delta J(\vec{x}', t_2)\delta J(\vec{x}, t_1)} < 0|U(T)|0 > . \quad (7.21)$$

Since we assumed that $|0>$ is an eigenstate of H (i.e., at $J = 0$), $< 0|U_{J=0}(T) = < 0|e^{-iE_0T}$ (with E_0 the vacuum energy) and we can bring the trivial phase factor e^{-iE_0T} to the other side by normalising with $< 0|U_{J=0}(T)|0 >$, as in Equation (7.17).

$$< 0|\hat{\varphi}(\vec{x}', t_2)\hat{\varphi}(\vec{x}, t_1)|0 >_{J=0}= \left[- < 0|U(T)|0 >^{-1} \frac{\delta^2 < 0|U(T)|0 >}{\delta J(\vec{x}', t_2)\delta J(\vec{x}, t_1)}\right]_{J=0},$$

$$(7.22)$$

where in the path integral formulation one has

$$< 0|U(T)|0 > = \int \prod_{\vec{p}} d\tilde{\varphi}(\vec{p})d\tilde{\varphi}'(\vec{p}) < 0|\{\tilde{\varphi}'(\vec{p})\} >$$

$$\times \int \mathcal{D}\varphi(x)\exp(i\int_0^T \mathcal{L}\, d_4x) < \{\tilde{\varphi}(\vec{p})\}|0 >, \quad (7.23)$$

with the boundary conditions as listed below Equation (7.20).

To study the role the source plays in creating and annihilating particles, we will calculate both in the Hamiltonian and in the path integral formulations the matrix element

$$< \vec{p}|U(T)|0 > , \quad |\vec{p} >\equiv a^\dagger(\vec{p})|0 > . \quad (7.24)$$

Hence $|\vec{p} >$ is the one-particle state with momentum $\vec{p}$. Using the result of Equations (5.7) and (5.8), which is equivalent to the result $|\Psi(t) >= U(t)|0 >$, we find for this matrix element in lowest nontrivial order

$$< \vec{p}|U(T)|0 > = -i\int_0^T dt\ < 0|a(\vec{p})e^{i(t-T)H_0}H_1(t)e^{-itH_0}|0 >$$

$$= -i\sum_{\vec{k}}\int_0^T dt\ < 0|a(\vec{p})e^{i(p_0(\vec{p})+E_0)(t-T)}\tilde{J}(\vec{k}, t)$$

$$\times \frac{(a(-\vec{k}) + a^\dagger(\vec{k}))}{\sqrt{2k_0(\vec{k})}}e^{-iE_0t}|0 >$$

$$= -i\frac{e^{-i(E_0+p_0(\vec{p}))T}}{\sqrt{2p_0(\vec{p})}}\int_0^T dt\ \tilde{J}(\vec{p}, t)e^{ip_0(\vec{p})t}$$

$$= -i\frac{\sqrt{\pi}e^{-i(E_0+p_0(\vec{p}))T}}{\sqrt{p_0(\vec{p})}}\tilde{J}(p). \quad (7.25)$$

To write down the path integral result, we first express the annihilation operator in terms of the field. From Equation (5.3) we find

$$a(\vec{p}) + a^{\dagger}(-\vec{p}) = \sqrt{\frac{2p_0(\vec{p})}{V}} \int d_3\vec{x}\, \hat{\varphi}(\vec{x}) e^{-i\vec{p}\cdot\vec{x}} \quad , \tag{7.26}$$

such that $\left[\text{using} < 0|a^{\dagger}(-\vec{p}) = 0\right]$

$$< \vec{p}|U(T)|0> = \sqrt{\frac{2p_0(\vec{p})}{V}} \int d_3\vec{x}\, e^{-i\vec{p}\cdot\vec{x}} < 0|\hat{\varphi}(\vec{x})U(T)|0> . \tag{7.27}$$

We now use Equations (7.14), (7.19) and an obvious generalisation of Equation (7.21), such that

$$< 0|\hat{\varphi}(\vec{x})U(T)|0> = < 0|U(T)\hat{\varphi}(\vec{x}, T)|0> = i\frac{\delta}{\delta J(\vec{x}, T)} < 0|U(T)|0>$$

$$= ie^{-iE_0 T} \frac{\delta}{\delta J(\vec{x}, T)} \exp\left(-\frac{i}{2} \int d_4 x d_4 y\, J(x)G(x-y)J(y)\right). \tag{7.28}$$

We evaluate this to linear order in the source J, using that in a finite volume the Green's function is given by

$$G(x-y) = \frac{1}{V} \sum_{\vec{k}} \int \frac{dk_0}{2\pi} \frac{e^{-ik(x-y)}}{k^2 - m^2 + i\varepsilon} \quad , \tag{7.29}$$

such that

$$< \vec{p}|U(T)|0> = e^{-iE_0 T} \sqrt{\frac{2p_0(\vec{p})}{V}} \int d_3\vec{x}\, e^{-i\vec{p}\cdot\vec{x}} \int d_4 y\, G(x-y)J(y) + \mathcal{O}(J^3)$$

$$= e^{-iE_0 T} \sqrt{\frac{p_0(\vec{p})}{\pi}} \int dp_0 \frac{\tilde{J}(p)e^{-ip_0 T}}{p^2 - m^2 + i\varepsilon} + \mathcal{O}(J^3). \tag{7.30}$$

Note that in the last step we integrate over p_0 as a dummy variable, which in the expression for the Green's function above is called k_0—this renaming is just for ease of notation. Also, $x_0 = T$ is assumed. For the p_0 integration we need the analytic behaviour of $\tilde{J}(p)$ for imaginary p_0 in order to see if we are allowed to deform the integration contour such that only one of the poles in the integrand contributes. Since $\tilde{J}(p) = \int_0^T dt\, e^{ip_0 t}\tilde{J}(\vec{p}, t)/\sqrt{2\pi}$, $\tilde{J}(p)$ will vanish for $\text{Im}\,p_0 \to \infty$, whereas $e^{-ip_0 T}\tilde{J}(p)$ will vanish for $\text{Im}\,p_0 \to -\infty$. In Equation (7.30) the p_0 integration can therefore be deformed to the lower half-plane in a clockwise fashion giving a minus sign and a residue from the pole at $p_0 = p_0(\vec{p}) \equiv \sqrt{\vec{p}^2 + m^2}$, which yields the result

$$< \vec{p}|U(T)|0 > = -i\frac{\sqrt{\pi}e^{-i(E_0+p_0(\vec{p}))T}}{\sqrt{p_0(\vec{p})}}\tilde{J}(p)$$

$$\times \exp\left(-\frac{i}{2}\int d_4x d_4y\, J(x)G(x-y)J(y)\right). \quad (7.31)$$

To linear order in the source J this coincides with the result of Equation (7.25). Again, the path integral trivially allows an extension to arbitrary order in the source, as indicated.

For later use, we will also consider the matrix element

$$< 0|U(T)|\vec{p} > = \sqrt{\frac{2p_0(\vec{p})}{V}}\int d_3\vec{x}\, e^{i\vec{p}\cdot\vec{x}} < 0|U(T)\hat{\varphi}(\vec{x},0)|0 >$$

$$= i\sqrt{\frac{2p_0(\vec{p})}{V}}\int d_3\vec{x}\, e^{i\vec{p}\cdot\vec{x}}\frac{\delta < 0|U(T)|0 >}{\delta J(\vec{x},0)}. \quad (7.32)$$

The analogue of Equation (7.30) becomes

$$< 0|U(T)|\vec{p} > = e^{-iE_0T}\sqrt{\frac{2p_0(\vec{p})}{V}}\int d_3\vec{x}\, e^{i\vec{p}\cdot\vec{x}}\int d_4y\, G(x-y)J(y) + \mathcal{O}(J^3)$$

$$= e^{-iE_0T}\sqrt{\frac{p_0(\vec{p})}{\pi}}\int dp_0\frac{\tilde{J}(-\vec{p},p_0)}{p^2-m^2+i\varepsilon} + \mathcal{O}(J^3), \quad (7.33)$$

in which case $x_0 = 0$ is assumed. Now we must deform the contour for the p_0 integration to the upper half-plane in a counterclockwise fashion such that the residue at the pole $p_0 = -p_0(\vec{p})$ contibutes. This gives, analogously to Equation (7.31),

$$< 0|U(T)|\vec{p} > = -i\frac{\sqrt{\pi}e^{-iE_0T}}{\sqrt{p_0(\vec{p})}}\tilde{J}(-p)\exp\left(-\frac{i}{2}\int d_4x d_4y\, J(x)G(x-y)J(y)\right).$$

$$(7.34)$$

Apart from the trivial difference of the factor $\exp(-ip_0T)$, we see that the amplitude for the annihilation of a one-particle state is proportional to $\tilde{J}(p)$ whereas it is proportional (exactly with the same factor) to $\tilde{J}(-p)$ for the creation of a one-particle state $\left[\text{in both cases } p_0 = p_0(\vec{p})\right]$.

8

Perturbative Expansion in Field Theory

DOI: 10.1201/b15364-8

As we have seen in the previous chapter $\left[\int \mathcal{D}\varphi(x)\right.$ where relevant includes ground-state factors$\left.\right]$

$$Z(J, g_n) \equiv < 0|U(T)|0 > = \int \mathcal{D}\varphi(x) \, \exp\left(i \int d_4x \, \mathcal{L}(\varphi)\right) \qquad (8.1)$$

will play the role of a generating functional for calculating expectation values of products of field operators, which will now be studied in more detail. In general the Lagrange density for a scalar field theory is given by

$$\mathcal{L}(\varphi) = \mathcal{L}_2(\varphi) - V(\varphi) - J(x)\varphi(x), \qquad (8.2)$$

where $\mathcal{L}_2(\varphi)$ is quadratic in the fields, hence for a scalar field

$$\mathcal{L}_2(\varphi) = \tfrac{1}{2}\left(\partial_\mu\varphi(x)\partial^\mu\varphi(x) - m^2\varphi^2(x)\right),$$

$$V(\varphi) = \frac{g_3}{3!}\varphi^3(x) + \frac{g_4}{4!}\varphi^4(x) + \cdots. \qquad (8.3)$$

As mentioned before, it is customary to not include the mass term in the potential V, such that V describes the interactions. We can add the interaction as an operator, when evaluating the path integral for the quadratic approximation $\mathcal{L}(\varphi) = \mathcal{L}_2(\varphi) - \varphi(x)J(x)$,

$$Z(J, g_n) = \int \mathcal{D}\varphi(x) \, \exp\left(i \int d_4x \, \{\mathcal{L}_2(\varphi) - \varphi(x)J(x)\}\right) \exp\left(-i \int d_4x \, V(\varphi)\right)$$

$$= < 0|U_2(T)T\exp\left(-i \int d_4x \, V(\hat{\varphi})\right)|0 >.$$

$$(8.4)$$

We can now use the fact that

$$\int \mathcal{D}\varphi(x) \prod_j \varphi(x_{(j)}) \exp\left(i \int d_4x(\mathcal{L}_2(\varphi) - \varphi(x)J(x))\right)$$

$$= \prod_j \left(\frac{i\delta}{\delta J(x_{(j)})}\right) \int \mathcal{D}\varphi(x) \exp\left(i \int d_4x\,(\mathcal{L}_2(\varphi) - \varphi(x)J(x))\right)$$

$$= \prod_j \left(\frac{i\delta}{\delta J(x_{(j)})}\right) \exp\left(-\frac{i}{2}\int d_4x d_4y\, J(x)G(x-y)J(y)\right), \quad (8.5)$$

to find a somewhat formal, but in an expansion with respect to the coupling constants g_n, well-defined expression for the fully interacting path integral

$$Z(J, g_n) = \exp\left(-i \int d_4x\, V\left(\frac{i\delta}{\delta J(x)}\right)\right)$$

$$\times \exp\left(-\frac{i}{2}\int d_4x d_4y\, J(x)G(x-y)J(y)\right)$$

$$\equiv \exp\left(-i \int d_4x\, V\left(\frac{i\delta}{\delta J(x)}\right)\right) Z_2(J). \quad (8.6)$$

We have assumed the vacuum energy to be normalised to zero, in absence of interactions, such that $Z(J = g_n = 0) = 1$. Equivalently, $Z(J, g_n)$ is synonymous with $Z(J, g_n)/Z(J = g_n = 0)$. We now define G_J as

$$G_J \equiv \log Z(J, g_n), \quad (8.7)$$

where the dependence on the coupling constants in G_J is implicit. We will show that G_J can be seen as the sum of all connected diagrams. A diagram is connected if it cannot be decomposed in the product of two diagrams that are not connected. Note that at $J = 0, iG_J/T$ equals the energy of the ground state as a function of the coupling constants, normalised so as to vanish at zero couplings.

The different diagrams arise from the expansion of

$$\exp\left(-i \int d_4x\, V\left(\frac{i\delta}{\delta J(x)}\right)\right) = \exp\left(-i \int d_4x \sum_{\ell=3} \frac{g_\ell}{\ell!}\left(\frac{i\delta}{\delta J(x)}\right)^\ell\right) \quad (8.8)$$

in powers of g_ℓ. Each factor $\frac{g_\ell}{\ell!}\left[i\delta/\delta J(x)\right]^\ell$ will represent an ℓ-point vertex, with coordinate x, which is to be integrated over. As we saw in the derivation of the classical equations of motion, the integral over x in the Fourier representation gives rise to conservation of momentum at the vertex. Using

$$\varphi(x) = \frac{1}{(2\pi)^2}\int d_4k\, e^{-ikx}\tilde{\varphi}(k), \quad J(x) = \frac{1}{(2\pi)^2}\int d_4k\, e^{-ikx}\tilde{J}(k), \quad (8.9)$$

TABLE 8.1

Feynman rules for scalars.

Coordinate space		Momentum space		
	$\equiv i^{\ell-1}g_\ell \int d_4x$		$\equiv i^{\ell-1}(2\pi)^{4-2\ell}g_\ell\delta_4(\sum_i k_i)$	vertex
	$\equiv -iG(x-y)$		$\equiv \int d_4k \frac{-i}{k^2-m^2+i\varepsilon}$	propagator
	$\equiv \int d_4x J(x)$		$\equiv \tilde{J}(k)$	source

we have for each vertex in the Fourier representation

$$-ig_\ell \int d_4x \left(\frac{i\delta}{\delta J(x)}\right)^\ell = -ig_\ell(2\pi)^4\delta_4\left(\sum_{j=1}^\ell k_{(j)}\right) \int \prod_{j=1}^\ell \left(\frac{d_4k_{(j)}}{(2\pi)^2}\frac{i\delta}{\delta\tilde{J}(k_{(j)})}\right).$$

(8.10)

Note that factors of 2π are dropping out in the identities

$$\exp\left(-\frac{i}{2}\int d_4x d_4y\, J(x)G(x-y)J(y)\right) = \exp\left(-\frac{i}{2}\int d_4k \frac{\tilde{J}(k)\tilde{J}(-k)}{k^2-m^2+i\varepsilon}\right),$$

$$\int d_4x\, \varphi(x)J(x) = \int d_4k\, \tilde{\varphi}(k)\tilde{J}(-k).$$

(8.11)

In the quantum theory we have to keep track of the factors i. Compared to the Feynman rules of Chapter 4, the propagator will come with an extra factor $-i$. A vertex will now carry a factor $ig_\ell\left[(2\pi)^2/i\right]^{2-\ell}$ (in a finite volume this becomes $ig_\ell\left[\sqrt{2\pi V/i}\right)^{2-\ell}\right]$); see Table 8.1. To compute the vacuum energy, there is an overall factor i since $E_0T = iG_{J=0} = i\log Z(J=0, g_n)$. The same factor of i applies for using the tree-level diagrams to solve the classical equations of motion. It is easy to see that these Feynman rules give identical results for these tree-level diagrams, as compared to the Feynman rules introduced in Chapter 4. The factors of i exactly cancel each other.

We note that the propagator connected to a source comes down whenever a derivative in the source acts on $Z_2(J)$; see Equation (8.7). When this derivative acts on terms that have already come down from previous derivatives, one of the sources connected to a propagator is removed and this connects that propagator to the vertex associated to $\delta/\delta J(x)$. As any derivative is connected to a vertex, the propagator either runs between a vertex and a source , between two vertices , or it connects two legs of the *same* vertex . The possibility of closed loops did not occur in solving the classical equations of motion and is specific to the quantum theory.

To prove that G_J only contains connected diagrams we write

$$G_J = \log \left\{ \exp \left(X(i\delta/\delta J) \right) \exp \left(Y(J) \right) \right\} 1, \tag{8.12}$$

with

$$X(i\delta/\delta J(x)) \equiv -i \int d_4x \, V(\delta/\delta J(x)),$$

$$Y(J) \equiv -\frac{i}{2} \int d_4x d_4y J(x) G(x-y) J(y). \tag{8.13}$$

As $J(x)$ and $\delta/\delta J(x)$ form an algebra (similar to the algebra of $\hat{x}$ and $\hat{p}$ in quantum mechanics, however generalised to infinite dimensions), also X and Y are elements from the algebra, and we can express G_J in a sum of multiple commutators using the Campbell–Baker–Hausdorff formula [see Equation (6.44)].

$$G_J = \left(X + Y + \tfrac{1}{2}[X, Y] + \frac{1}{12}\left[X, [X, Y]\right] + \frac{1}{12}\left[Y, [Y, X]\right] + \cdots \right) 1. \tag{8.14}$$

Due to the multiple commutators, all components that do not commute are connected. However, if the components would commute they would not contribute to the commutators. This is even true if we do not put the derivatives with respect to the source to zero, once they have been moved to the right (this is why we consider the action on the identity).

The only thing that remains to be discussed is with which combinatorial factor each diagram should contribute. This is, as in Chapter 4, with the inverse of the order of the permutation group that leaves the topology of the diagram unchanged. These combinatorial factors are clearly independent of the space-time integrations and possible contractions of vector or other indices. We can check them by reducing the path integral to zero dimensions, or $\varphi(x) \to \varphi$ and $\mathcal{D}\varphi(x) \to d\varphi$. In other words, we replace the path integral by an ordinary integral. As an example consider

$$Z(J, g) = C \int d\varphi \, \exp \left(i \left\{ \tfrac{1}{2}\varphi M \varphi - \frac{g}{3!}\varphi^3 - \varphi J \right\} \right)$$

$$= \exp \left(-i\frac{g}{3!} \left(\frac{i\partial}{\partial J} \right)^3 \right) \exp \left(-\frac{i}{2} J M^{-1} J \right). \tag{8.15}$$

The constant C is simply to normalise $Z(J = g_n = 0) = 1$. Expanding the exponents we get in lowest nontrivial order

$$Z(J = 0) = 1 - \frac{g^2}{2(3!)^3} \left(\frac{i\partial}{\partial J} \right)^6 \left(-\frac{i}{2} J M^{-1} J \right)^3 \cdots$$

$$= \exp \left(i \frac{g^2}{2^4(3!)^3} \left(\frac{\partial}{\partial J} \right)^6 (J M^{-1} J)^3 + \mathcal{O}(g^3) \right)$$

$$= \exp \left(i \frac{5g^2}{24 M^3} + \mathcal{O}(g^3) \right)$$

$$= \exp \left(\frac{1}{12} \ominus + \frac{1}{8} \bigcirc\!-\!\bigcirc + \mathcal{O}(g^3) \right). \qquad (8.16)$$

In the last term, the numerical factors in front of the diagrams indicate the combinatorial factors (for the first diagram a factor 2 from interchanging the two vertices and a factor 3! from interchanging the three propagators; for the second diagram the latter factor is replaced by 4 as we can only interchange for each vertex the two legs that do not interconnect the two vertices). The Feynman rules for this simple case are that each vertex gets a factor $-g$ (in zero dimensions there are no factors 2π) and each propagator gets a factor $-i/M$. In Problem 13 the exponentiation is checked for $Z(J)$ to $\mathcal{O}(g^2)$ and $\mathcal{O}(J^5)$ (giving the simplest nontrivial check).

We will now show how the number of loops in a diagram is related to the expansion in $\hbar$. We can expect such a relation, as we have shown at the end of Chapter 6 that the $\hbar \to 0$ limit is related to the classical equations of motion, whereas we have shown in Chapter 4 that these classical equations are solved by tree diagrams. If we call L the number of loops of a diagram, we will show that

$$Z(J, g_n) = \int \mathcal{D}\varphi \, \exp \left(\frac{i}{\hbar} \int (\mathcal{L}(\varphi) - \varphi J) \right) \equiv \exp(G_J/\hbar), \, G_J = \sum_{L=0}^{\infty} \hbar^L G_{L,J},$$

$$(8.17)$$

where $G_{L,J}$ is the sum of all connected diagrams with exactly L loops. This means that a loop expansion is equivalent with an expansion in $\hbar$. To prove this we first note that due to reinstating $\hbar$ the source term will get an extra factor $1/\hbar$, the propagator a factor $\hbar$, and the coupling constants g_n are replaced by $g_n/\hbar$. A diagram with V_n n-point vertices, E external lines (connected to a source), and P propagators has therefore an extra overall factor of

$$\hbar^{-E} \hbar^P \prod_{n=3} \hbar^{-V_n}. \qquad (8.18)$$

We can relate this to the number of loops by noting that the number of momentum integrations (i.e., the number of independent momenta) in a diagram equals the number of loops plus the number of external lines, minus one for the overall conservation of energy and momentum, i.e., $L + E - 1$. On the other hand, the number of momentum integrations is also the number of propagators minus the number of delta functions coming from the vertices, i.e., $P - \sum_{n=3} V_n$. Hence

$$L = 1 + P - E - \sum_{n=3} V_n,\qquad(8.19)$$

which implies that the total number of $\hbar$ factors in a diagram is given by $L-1$. In the next chapters we will often consider so-called amputated diagrams, where the external propagators connected to a source are taken off from the expressions for the diagram. If we do not count these external propagators, Equation (8.19) has to be replaced by $L = 1 + P - \sum_{n=3} V_n$, as there are exactly E such external propagators.

9

The Scattering Matrix

DOI: 10.1201/b15364-9

We would like to compute the amplitude for the transition of n incoming particles at $t = T_{in}$ to ℓ outgoing particles at $t = T_{out}$ in the limit where $T_{out} \to \infty$ and $T_{in} \to -\infty$. The difference with quantum mechanics is that the particle number is no longer conserved.

$$_{out}< \vec{p}_1, \vec{p}_2, \ldots, \vec{p}_\ell | \vec{k}_1, \vec{k}_2, \ldots, \vec{k}_n >_{in}$$

$$\equiv < \vec{p}_1, \vec{p}_2, \ldots, \vec{p}_\ell | U(T_{out}, T_{in}) | \vec{k}_1, \vec{k}_2, \ldots, \vec{k}_n >. \tag{9.1}$$

In terms of creation and annihilation operators this can be written as

$$_{out}< \vec{p}_1, \vec{p}_2, \ldots, \vec{p}_\ell | \vec{k}_1, \vec{k}_2, \ldots, \vec{k}_n >_{in}$$

$$= < 0|a(\vec{p}_1)a(\vec{p}_2)\cdots a(\vec{p}_\ell)U(T_{out}, T_{in})a^\dagger(\vec{k}_1)a^\dagger(\vec{k}_2)\cdots a^\dagger(\vec{k}_n)|0 >. \tag{9.2}$$

From Equations (7.26), (7.28), and (7.32) we know how to implement these creation and annihilation operators on the generating functional $Z(J, g_n)$

$$\hat{a}^\dagger(\vec{p}) = i\sqrt{\frac{2p_0(\vec{p})}{V}} \int d_3\vec{x} \; e^{i\vec{p}\cdot\vec{x}} \frac{\delta}{\delta J(\vec{x}, t = T_{in})} = i\sqrt{2p_0(\vec{p})} \frac{\delta}{\delta \tilde{J}(\vec{p}, t = T_{in})},$$

$$\hat{a}(\vec{p}) = i\sqrt{\frac{2p_0(\vec{p})}{V}} \int d_3\vec{x} \; e^{-i\vec{p}\cdot\vec{x}} \frac{\delta}{\delta J(\vec{x}, t = T_{out})} = i\sqrt{2p_0(\vec{p})} \frac{\delta}{\delta \tilde{J}(-\vec{p}, t = T_{out})}. \tag{9.3}$$

This implies the following identity for the scattering matrix

$$_{out}< \vec{p}_1, \vec{p}_2, \ldots, \vec{p}_\ell | \vec{k}_1, \vec{k}_2, \ldots, \vec{k}_n >_{in} = \prod_{i=1}^{\ell} \hat{a}(\vec{p}_i) \prod_{j=1}^{n} \hat{a}^\dagger(\vec{k}_j) \exp(G_J)|_{J=0}. \tag{9.4}$$

In principle, this allows us to calculate the scattering, taking $T_{in} \to -\infty$ and $T_{out} \to \infty$.

There is, however, a problem to associate the particle states in the presence of interactions with the ones we have derived from the noninteracting theory. The problem is that particles can have self-interactions long before and after the different particles have scattered off each other. We have to reconsider our notion of particle states, as in experiments we are unable to switch off these self-interactions. For simplicity we assume that the one-particle states are stable, as in the simple scalar theory we have been considering. This implies from conservation of probability that

$$\int d_3\vec{k} \mid _{\text{out}}< \vec{p}|\vec{k}>_{\text{in}} \mid^2 = 1, \tag{9.5}$$

independently of $\vec{p}$. In general, conservation of probability implies that the S-matrix is unitary. Formally, unitarity of an S-matrix is guaranteed as soon as the Hamiltonian is Hermitian. Because of the necessity to regulate the quantum theory, e.g., by introducing a cutoff, this is generally no longer true and one has to show that unitarity is restored when the cutoff is removed. If this is not possible, the theory is ill-defined or at best does not make sense above the energies where unitarity is violated.

For the free theory, unitarity is of course satisfied. In this case, the only diagram contributing to G_J is the one with a single propagator connecting two sources, which is also called the connected two-point function $G_c^{(2)}(J)$

$$G_J = G_c^{(2)}(J) = -\frac{i}{2} \int \int dt ds \sum_{\vec{p}} \int \frac{dp_0}{2\pi} e^{-ip_0(t-s)} \frac{\tilde{J}(\vec{p}, t)\tilde{J}(-\vec{p}, s)}{p^2 - m^2 + i\varepsilon}. \tag{9.6}$$

This implies that ($T \equiv T_{\text{out}} - T_{\text{in}}$)

$$_{\text{out}} < \vec{p}|\vec{k} >_{\text{in}} = \hat{a}(\vec{p})\hat{a}^\dagger(\vec{k}) \exp\left(G_c^{(2)}(J)\right)|_{J=0} = 2ip_0(\vec{p})\delta_{\vec{k},\vec{p}} \int \frac{dp_0}{2\pi} \frac{e^{ip_0 T}}{p^2 - m^2 + i\varepsilon}$$

$$= e^{-ip_0(\vec{p})T} \delta_{\vec{k},\vec{p}}, \tag{9.7}$$

where the p_0 integration is performed by deforming the contour to the upper half-plane, giving a contribution from the pole at $p_0 = -p_0(\vec{p})$ only. It is trivial to see that this is the same as what can be obtained within the Hamiltonian formulation.

In the presence of interactions, this result is no longer true, since the connected two-point function will deviate from the one in the free theory. In this case we can write (from now on the symmetry factors will be absorbed in the expression associated to a diagram)

$$\tag{9.8}$$

where we have written the connected two-point function in terms of the one-particle irreducible (1PI) two-point function $i\Sigma(p)$ (Σ is the so-called self-energy)

$$i\Sigma(p) \equiv \tag{9.9}$$

In general, a $1PI$-graph is a connected graph that remains connected when one arbitrary propagator is being cut (*except* when cutting away a tadpole of the form ⊸◯, which we will *not* allow. However, these tadpoles describe single particles popping in or out of the vacuum, usually required to be absent. They can be removed by shifts in the fields.). The external lines of these diagrams will carry no propagator. The diagrams of Equation (9.8) can be converted to the result

$$
\frac{1}{2} \int d_4 p \, |\tilde{J}(p)|^2 \left\{ \frac{-i}{p^2 - m^2 + i\varepsilon} + \left(\frac{-i}{p^2 - m^2 + i\varepsilon} \right)^2 i\Sigma(p) \right.
$$

$$
\left. + \left(\frac{-i}{p^2 - m^2 + i\varepsilon} \right)^3 \left(i\Sigma(p) \right)^2 + \cdots \right\}
$$

$$
= \frac{1}{2} \int d_4 p \, |\tilde{J}(p)|^2 \left\{ \frac{-i}{p^2 - m^2 + i\varepsilon} \sum_{n=0}^{\infty} \left(\frac{\Sigma(p)}{p^2 - m^2 + i\varepsilon} \right)^n \right\}
$$

$$
= -\frac{i}{2} \int d_4 p \, \frac{|\tilde{J}(p)|^2}{p^2 - m^2 - \Sigma(p) + i\varepsilon} \equiv G_c^{(2)}(J). \tag{9.10}
$$

Normally the self-energy will not vanish at $p^2 = m^2$, such that the self-interactions shift the pole in the two-point function to another value, $\tilde{m}^2$, i.e.,

$$
p^2 - m^2 - \Sigma(p) = 0 \quad \text{for} \quad p_0^2 = \vec{p}^2 + \tilde{m}^2. \tag{9.11}
$$

Consequently, the mass of the one-particle states is shifted (or *renormalised*). As we cannot switch off the interactions in nature, the *true* or observable mass is $\tilde{m}$ and not m; the latter is also called the bare mass. The residue at the poles (i.e., $p^2 = \tilde{m}^2$, called the mass-shell) will in general also change from $\pm\pi i / p_0(\vec{p})$ to $\pm\pi i Z / p_0(\vec{p})$. On the mass-shell [i.e., $\tilde{J}(p)$ vanishes rapidly as a function of $|p^2 - \tilde{m}^2|$] one therefore has

$$
G_c^{(2)}(J) = -\frac{i}{2} \int d_4 p \, \frac{Z |\tilde{J}(p)|^2}{p^2 - \tilde{m}^2 + i\varepsilon}. \tag{9.12}
$$

As long as the one-particle states are stable, Equation (9.5) needs to remain valid, which can only be achieved [see Equation (9.7)] by rescaling the wave functionals with a factor $\sqrt{Z}$ (this is called *wavefunction renormalisation*). It implies that Equation (9.4) needs to be modified to

$$
{}_{\text{out}}< \vec{p}_1, \vec{p}_2, \ldots, \vec{p}_\ell | \vec{k}_1, \vec{k}_2, \ldots, \vec{k}_n >_{\text{in}} = \prod_{i=1}^{\ell} \hat{a}_-(\vec{p}_i) \prod_{j=1}^{n} \hat{a}_+(\vec{k}_j) \exp(G_J)|_{J=0},
$$

$$
\tag{9.13}
$$

where

$$\hat{a}_+(\vec{p}) \equiv i \left(\frac{2\sqrt{\vec{p}^2 + \tilde{m}^2}}{Z} \right)^{\frac{1}{2}} \frac{\delta}{\delta \tilde{J}(\vec{p}, t = T_{\text{in}})},$$

$$\hat{a}_-(\vec{p}) \equiv i \left(\frac{2\sqrt{\vec{p}^2 + \tilde{m}^2}}{Z} \right)^{\frac{1}{2}} \frac{\delta}{\delta \tilde{J}(-\vec{p}, t = T_{\text{out}})}.$$

(9.14)

As for the free theory, each of these operators $\hat{a}_\pm(\vec{p})$ will replace one external line (propagator plus source) by an appropriate wave-function factor and puts these external lines on the mass-shell. We will call the connected n-point function with amputated external lines the amputated connected n-point function $G_c^{(\text{amp})}(p_1, p_2, \ldots, p_n)$ (in general not one-particle irreducible), i.e.,

$$G_c^{(n)}(J) \equiv \prod_{i=1}^{n} \left\{ \int d_4 p_i \; \frac{-i\tilde{J}(p_i)}{p_i^2 - m^2 - \Sigma(p_i) + i\varepsilon} \right\} G_c^{(\text{amp})}(p_1, p_2, \ldots, p_n). \quad (9.15)$$

Diagrammatically this looks as follows:

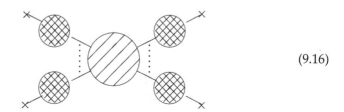

(9.16)

To get the S-matrix we have to compute [see Equation (9.13)]

$$\left(\prod_{i=1}^{\ell} \hat{a}_-(\vec{p}_i) \prod_{j=1}^{n} \hat{a}_+(\vec{k}_j) \, e^{G_J} \right)_{J=0} = e^{G_{J=0}} \prod_{i=1}^{\ell} \hat{a}_-(\vec{p}_i)$$

$$\times \prod_{j=1}^{n} \hat{a}_+(\vec{k}_j) \left(\sum_{\Sigma r q_r = \ell + n} \prod_r \frac{1}{q_r!} \{G_c^{(r)}(J)\}^{q_r} \right).$$

(9.17)

The sum is over all possible partitions of $\ell + n$. Let us first consider the most important term, corresponding to connected graphs, where $q_1 = \ell + n$

$$\prod_{i=1}^{\ell} \hat{a}_-(\vec{p}_i) \prod_{j=1}^{n} \hat{a}_+(\vec{k}_j)\, G_c^{(\ell+n)}(J)$$

$$= \prod_{j=1}^{n} -i\sqrt{\pi Z/k_0^{(j)}} \prod_{i=1}^{\ell} -i\sqrt{\pi Z/p_0^{(i)}}\, G_c^{(\mathrm{amp})}(\{-p_i\}, \{k_j\})e^{-i\Sigma p_0^{(i)} T}$$

$$\equiv \frac{-2\pi i V \delta_4\left(\sum_{i=1}^{\ell} p_i - \sum_{j=1}^{n} k_j\right) \mathcal{M}_{\ell+n}(\{-p_i\}, \{k_j\})}{\sqrt{\prod_{i=1}^{\ell} 2p_0^{(i)} V \prod_{j=1}^{n} 2k_0^{(j)} V}} e^{-i\Sigma p_0^{(i)} T}. \quad (9.18)$$

Here $\mathcal{M}_\ell$ (a Lorentz scalar as we will see later) is the so-called reduced matrix element with ℓ external lines, all on the mass-shell. Note that we have extracted the trivial energy factors [remember that $\sum p_0^{(i)} T_{\mathrm{out}} - \sum k_0^{(j)} T_{\mathrm{in}} = \sum p_0^{(i)}(T_{\mathrm{out}} - T_{\mathrm{in}}) \equiv \sum p_0^{(i)} T$], such that the limits $T_{\mathrm{in}} \to -\infty$ and $T_{\mathrm{out}} \to \infty$ can be taken. Each $\hat{a}(\vec{p})_\pm$ will act on one of the factors between curly brackets in Equation (9.15). Concentrating on one such a factor we have

$$\int d_4 p\, \frac{-i J(p) G_c^{(\mathrm{amp})}(p, p_2, \ldots, p_n)}{p^2 - m^2 - \Sigma(p) + i\varepsilon}$$

$$= \int \frac{d_4 p}{\sqrt{2\pi}} \int dt\, \frac{-i J(\vec{p}, t) e^{i p_0 t} G_c^{(\mathrm{amp})}(p, p_2, \ldots, p_n)}{p^2 - m^2 - \Sigma(p) + i\varepsilon}, \quad (9.19)$$

such that, using Equation (9.14)

$$\hat{a}_-(\vec{p}) \int d_4 p\, \frac{-i J(p) G_c^{(\mathrm{amp})}(p, p_2, \ldots, p_n)}{p^2 - m^2 - \Sigma(p) + i\varepsilon}$$

$$= \sqrt{\frac{p_0(\vec{p})}{\pi Z}} \int dp_0\, \frac{e^{i p_0 T_{\mathrm{out}}} G_c^{(\mathrm{amp})}((p_0, -\vec{p}), \ldots, p_n)}{p^2 - m^2 - \Sigma(p) + i\varepsilon}$$

$$= -i\sqrt{2\pi V Z}\, \frac{G_c^{(\mathrm{amp})}(-p, p_2, \ldots, p_n)}{\sqrt{2p_0(\vec{p}) V}} e^{-i p_0(\vec{p}) T_{\mathrm{out}}}. \quad (9.20)$$

Since $T_{\mathrm{out}} \to \infty$, we can extend the p_0 contour integration to the upper half-plane, under mild regularity conditions for $G_c^{(\mathrm{amp})}(p, p_2, \ldots, p_n)$ as $\mathrm{Im}\, p_0 \to -\infty$ (that can easily be shown to be satisfied at any finite order in perturbation theory). Thus, the integral over p_0 only gets a contribution from the pole at $p_0 = -p_0(\vec{p}) = -\sqrt{\vec{p}^2 + \tilde{m}^2}$, with residue $-Z/2p_0(\vec{p})$ [compare this to

Equation (9.12)]. For the creation operators one similarly finds

$$\hat{a}_+(\vec{p}) \int d_4 p \, \frac{-i \tilde{J}(p) G_c^{(\mathrm{amp})}(p, p_2, \ldots, p_n)}{p^2 - m^2 - \Sigma(p) + i\varepsilon}$$

$$= \sqrt{\frac{p_0(\vec{p})}{\pi Z}} \int dp_0 \, \frac{e^{ip_0 T_{\mathrm{in}}} G_c^{(\mathrm{amp})}((p_0, \vec{p}), \ldots, p_n)}{p^2 - m^2 - \Sigma(p) + i\varepsilon}$$

$$= -i\sqrt{2\pi V Z} \, \frac{G_c^{(\mathrm{amp})}(p, p_2, \ldots, p_n)}{\sqrt{2 p_0(\vec{p}) V}} e^{ip_0(\vec{p}) T_{\mathrm{in}}}. \tag{9.21}$$

Likewise, as $T_{\mathrm{in}} \to -\infty$, we can now extend the p_0 contour integration to the lower half-plane [under the same regularity conditions for the amputated n-point functions as for Equation (9.20) to be valid], such that we pick up the contribution of the pole at $p_0 = p_0(\vec{p})$ with residue $Z/2p_0(\vec{p})$. Combining these results proves the first identity in Equation (9.18); the second is merely a definition.

Note that the derivation is not valid for the case $\ell + n = 2$, where the generalisation of Equation (9.7) to the interacting case implies

$$_{\mathrm{out}}< \vec{p} | \vec{k} >_{\mathrm{in}} = \hat{a}_-(\vec{p}) \hat{a}_+(\vec{k}) \, \exp(G_J)|_{J=0} = \delta_3(\vec{k} - \vec{p}) e^{-i[E_0 + p_0(\vec{p})]T}. \tag{9.22}$$

Here we used Equation (9.12) and the fact that $\exp(G_{J=0}) = \exp(-i E_0 T)$, which is often also normalised to 1, but till now we had only required this to be the case at zero couplings. The reason this case is special is because Equation (9.15) requires us to define for the amputated two-point function

$$G_c^{(\mathrm{amp})}(p_1, p_2) \equiv i\delta_4(p_1 + p_2)(p_1^2 - m^2 - \Sigma(p_1) + i\varepsilon), \tag{9.23}$$

which vanishes on the mass-shell. In using Equation (9.20) and Equation (9.21) it was implicitly assumed that the amputated Green's function has no zero that will cancel the pole.

The Feynman rules in momentum space for computing the reduced matrix elements will obviously have to be modified for the external lines to a factor $-i\sqrt{2\pi V Z}$ and an overall factor $i/(2\pi V)$ [as always, in an infinite volume one replaces V by $(2\pi)^3$]. If we associate a momentum delta function to a vertex and a momentum integration to a propagator (as was done up to now), the delta function for overall energy and momentum conservation should not be written explicitly in Equation (9.18), since it is contained in the reduced matrix element. Instead, if we choose to integrate over the independent loop momenta, implementing energy and momentum conservation at each vertex (as will be done from now on), the definition of Equation (9.18) is the appropriate one. The overall factors of i, 2π and V can be determined with the help of the two identities

$$L = P + 1 - \sum_n V_n \quad , \quad E + 2P = \sum_n n V_n. \tag{9.24}$$

TABLE 9.1

Modified Feynman rules for scalars.

momentum space		Itzykson and Zuber		
$k_3 \swarrow^{k_1} \searrow k_2$	$\equiv$ g_ℓ and $\sum k_i = 0$	$k_3 \swarrow^{k_1} \searrow k_2$	$\equiv$ $-i(2\pi)^4 g_\ell \delta_4(\sum_i k_i)$	vertex
$\underset{k}{\rule{2cm}{0.4pt}}$	$\equiv$ $\dfrac{1}{k^2 - m^2 + i\varepsilon}$	$\underset{k}{\rule{2cm}{0.4pt}}$	$\equiv$ $\int \dfrac{d_4 k}{(2\pi)^4} \dfrac{i}{k^2 - m^2 + i\varepsilon}$	propagator
$\multimap$	$\equiv$ $\sqrt{Z}$	$\multimap$	$\equiv$ $\sqrt{Z}$	external line
	$i \int \dfrac{d_4 k}{(2\pi)^4}$		1	loop factor

The proof for the first identity was discussed below Equation (8.19). For the second identity we put a dot on each end of a propagator ($\overset{\bullet}{\rule{0.6cm}{0.4pt}}\overset{\bullet}{}$) and one dot on each external line ($\overset{\bullet}{\rule{0.6cm}{0.4pt}}$), giving a total of $2P + E$ dots. The same dots can also be associated to each line of a vertex ($\overset{}{\cdot\bigwedge}$), giving $\sum n V_n$ dots, thus proving the second identity (see also Problem 15). To keep the derivation general, we evaluate the overall factor in a finite volume

$$i^{-P} i^{-E+1} i^{\Sigma(n-1)V_n} (2\pi V)^{-\frac{1}{2}\Sigma(n-2)V_n} (2\pi V)^{\frac{1}{2}E-1} = \left(\frac{i}{2\pi V}\right)^L. \qquad (9.25)$$

This implies that we can shift all numerical factors from the propagators, vertices and external lines to a factor $i/(2\pi V)$ [or $i/(2\pi)^4$ in an infinite volume] for each loop, giving the Feynman rules listed in Table 9.1 for an infinite volume. Note that the extraction of the factor $-2\pi i V$ in the definition of $\mathcal{M}$ is merely a convention (such that in lowest order $\mathcal{M}_n = g_n$). In the literature many different conventions are being used. As an example, Table 9.1 compares our Feynman rules with those of Itzykson and Zuber. Their convention for $\mathcal{M}_n$ is likewise to make it coincide to lowest order with the n-point vertex. However, as the latter does already contain a factor $-i(2\pi)^4$ (in a finite volume $-2\pi i V$), that factor should be absent in relating the reduced matrix element to the amputated n-point function. Combining the extra factors of i and 2π in the Feynman rules of Itzykson and Zuber gives $i(2\pi)^{-4} i^{P-\Sigma V_n} (2\pi)^{4(\Sigma V_n - P)} = [i/(2\pi)^4]^L$, guaranteeing equivalence of the two sets of Feynman rules.

Concerning the symmetry factor associated to a particular diagram, we note the following. As we have generally fixed the external momenta, interchanging external lines is no longer allowed. But from Equation (9.15) we see that the symmetry factor $n!$, to be taken into account for $G_c^{(n)}(J)$, will be compensated by the n derivatives on n sources. Hence, in computing the reduced matrix elements, the symmetry factors are determined without allowing for permutations on the external lines.

To conclude this chapter we return to Equation (9.17) and discuss the contributions that will be associated to the diagrams that are not connected. Each factor of $q_r!$ is compensated for by the differentiations on $\{G_c^{(r)}(J)\}^{q_r}$. For the corresponding connected components, the rules are identical to the ones specified above. In particular each connected component will carry its own factor $-i(2\pi)^4\delta_4(\sum p_i)$ for the conservation of energy and momentum (p_i is now assumed to run over a subset of both the incoming and the negative of the outgoing four momenta). In a physical picture the disconnected parts correspond to situations where only a subset of the incoming particles will interact with each other (the ones connected by a particular diagram). Quite often, the experimental situation is such that the energy-momentum conservation will only be compatible with the fully connected part. We just have to avoid the incoming momenta to coincide with any of the outcoming momenta. In a collider, this means one excludes particles that escape in the direction of the beams, where indeed it is not possible to put a detector. As an illustration, we will give the situation for $n = \ell = 2$ and all momenta nonzero (to avoid tadpole diagrams, —◯) to second order in the three-point coupling g_3, putting all other couplings to zero

$$_{\text{out}} < \vec{p}_1, \vec{p}_2 | \vec{k}_1, \vec{k}_2 >_{\text{in}}$$

$$= \exp\left(-i[E_0 + p_0^{(1)} + p_0^{(2)}]T\right)\left[\delta_3(\vec{p}_1 - \vec{k}_1)\delta_3(\vec{p}_2 - \vec{k}_2)\left(\overline{\hspace{2cm}}\right)\right.$$

$$+ \delta_3(\vec{p}_1 - \vec{k}_2)\delta_3(\vec{p}_2 - \vec{k}_1)\left(\times\right)$$

$$-i\frac{(2\pi)^4\delta_4(p_1 + p_2 - k_1 - k_2)}{\sqrt{2p_0^{(1)}(2\pi)^3 2p_0^{(2)}(2\pi)^3 2k_0^{(1)}(2\pi)^3 2k_0^{(2)}(2\pi)^3}}$$

$$\left.\times\left(\overline{\hspace{1cm}} + \times + \mathcal{O}(g_3^3)\right)\right]. \tag{9.26}$$

The first two diagrams, which have to be treated with special care [see Equations (9.22) and (9.23)], represent the situation without scattering. By definition they have no higher-order corrections.

10

Cross Sections

DOI: 10.1201/b15364-10

In many experimental situations, we are interested in the scattering of two particles with momenta k_1 and k_2 to a state with n particles with momenta $p_1, p_2, \ldots, p_n$. We denote by $\int d_3\vec{k}_1 \Psi_1(\vec{k}_1)|\vec{k}_1 >$ and $\int d_3\vec{k}_2 \Psi_2(\vec{k}_2)|\vec{k}_2 >$ the wave functionals of the incoming particles. This is to describe the more realistic case of a wave packet. The amplitude for scattering to take place is hence given by

$$\mathcal{A} = -i(2\pi)^4 \int d_3\vec{k}_1 d_3\vec{k}_2 \frac{\delta_4\left(\sum_{i=1}^n p_i - k_1 - k_2\right)\mathcal{M}_{2+n}(\{-p_i\}, \{k_j\})}{\sqrt{\prod_{i=1}^n 2p_0^{(i)}(\vec{p}_i)(2\pi)^3 \prod_{j=1}^2 2k_0^{(j)}(\vec{k}_j)(2\pi)^3}}$$

$$\times \tilde{\Psi}_1(\vec{k}_1)\tilde{\Psi}_2(\vec{k}_2)e^{-i[E_0 + \Sigma p_0^{(i)}]T}. \tag{10.1}$$

If we define the wave function in coordinate space as usual

$$\Psi_j(x) = \int \frac{d_3\vec{k}}{\sqrt{2k_0(\vec{k})(2\pi)^3}} e^{-ikx}\tilde{\Psi}_j(\vec{k}), \tag{10.2}$$

we can compute the overlap of the two wave functions

$$\int d_4x \, \Psi_1(x)\Psi_2(x)e^{ipx} = (2\pi)^4 \int \frac{d_3\vec{k}_1}{\sqrt{2k_0(\vec{k}_1)(2\pi)^3}} \frac{d_3\vec{k}_2}{\sqrt{2k_0(\vec{k}_2)(2\pi)^3}}$$

$$\times \delta_4(k_1 + k_2 - p)\tilde{\Psi}_1(\vec{k}_1)\tilde{\Psi}_2(\vec{k}_2). \tag{10.3}$$

We assume that over the range of momenta in the wave packets, the reduced matrix elements are constant (which can be achieved with arbitrary precision for arbitrarily narrow wave packets in momentum space). This allows us to write for the scattering probability of two particles into n particles, with momenta in between p_i and $p_i + dp_i$,

$$dW = |\mathcal{M}(\{-p_i\}, \{\vec{k}_j\})|^2 f(p) \prod_{i=1}^n \frac{d_3\vec{p}_i}{2p_0(\vec{p}_i)(2\pi)^3},$$

$$f(p) \equiv \int d_4x d_4y \, \Psi_1(x)\Psi_2(x)\Psi_1^*(y)\Psi_2^*(y)e^{ip(x-y)}, \tag{10.4}$$

where $p = \sum_{i=1}^{n} p_i = k_1 + k_2$. The momenta $\bar{k}_i$ in the reduced matrix element are the central values of the wave packet in momentum space for the two incoming particle beams. Under the same assumption that the momentum spread in the beams is very small, the function $f(p)$ will be highly peaked around $p = \bar{k}_1 + \bar{k}_2$, such that

$$f(p) \approx \delta_4(p - \bar{k}_1 - \bar{k}_2) \int d_4 p \; f(p) = (2\pi)^4 \delta_4(p - \bar{k}_1 - \bar{k}_2)$$

$$\times \int d_4 x |\Psi_1(x)|^2 |\Psi_2(x)|^2. \qquad (10.5)$$

The quantities $|\Psi_j(x)|^2$ are of course related to the probability densities of the two particles in their respective beams,

$$\rho_j(x) \equiv i\left(\Psi_j^*(x)\partial_t \Psi_j(x) - \Psi_j(x)\partial_t \Psi_j^*(x)\right) \approx 2\bar{k}_0^{(j)}|\Psi_j(x)|^2, \qquad (10.6)$$

again using the fact that the wave packet is highly peaked in momentum space. Putting these results together we find

$$dW = (2\pi)^4 \delta_4\left(\sum p_i - \bar{k}_1 - \bar{k}_2\right) |\mathcal{M}(\{-p_i\}, \{\bar{k}_j\})|^2$$

$$\times \prod_{i=1}^{n} \frac{d_3 \vec{p}_i}{2p_0(\vec{p}_i)(2\pi)^3} \int d_4 x \frac{\rho_1(x)\rho_2(x)}{4\bar{k}_0^{(1)}\bar{k}_0^{(2)}}. \qquad (10.7)$$

Since $\rho_j(x)$ will depend on the experimental situation, we should normalise with respect to the total number of possible interactions in the experimental setup, also called the integrated luminosity L.

$$\int dt \, L(t) \equiv \int d_4 x \, \rho_1(x)\rho_2(x)|\vec{v}_1 - \vec{v}_2|. \qquad (10.8)$$

Here $\int d_3 \vec{x} \, \rho_1(x)\rho_2(x)$ is the number of possible interactions per unit volume at a given time and $|\vec{v}_1 - \vec{v}_2|$ is the relative velocity of the two beams. We have assumed that one of the velocities is zero (fixed target) or that the two velocities are parallel (colliding beams). Hence, $L(t) = \int d_3 \vec{x} \, \rho_1(\vec{x}, t)\rho_2(\vec{x}, t)|\vec{v}_1 - \vec{v}_2|$ is a flux, typically of the order of $10^{28} - 10^{33} \text{cm}^{-2}\text{s}^{-1}$. To consider the general case we note that we can also write

$$\int dt \, L(t) = \sqrt{(\bar{k}_1 \cdot \bar{k}_2)^2 - m_1^2 m_2^2} \int d_4 x \frac{\rho_1(x)\rho_2(x)}{\bar{k}_0^{(1)}\bar{k}_0^{(2)}}. \qquad (10.9)$$

After all, for a fixed target situation $\bar{k}_2 = (m_2, \vec{0})$, such that

$$\frac{\sqrt{(\bar{k}_1 \cdot \bar{k}_2)^2 - m_1^2 m_2^2}}{\bar{k}_0^{(1)}\bar{k}_0^{(2)}} = \frac{\sqrt{(\bar{k}_0^{(1)})^2 - m_1^2}}{\bar{k}_0^{(1)}} = |\vec{v}_1|, \qquad (10.10)$$

whereas for colliding beams of particles and antiparticles with mass m, where $\bar{k}_1 = (E, \vec{k}) = E(1, \vec{v})$ and $\bar{k}_2 = (E, -\vec{k}) = E(1, -\vec{v})$, one finds

$$\frac{\sqrt{(\bar{k}_1 \cdot \bar{k}_2)^2 - m_1^2 m_2^2}}{\bar{k}_0^{(1)} \bar{k}_0^{(2)}} = \frac{\sqrt{(E^2 + \vec{k}^2)^2 - m^4}}{E^2} = 2|\vec{k}|/E = 2|\vec{v}|, \quad (10.11)$$

such that both expressions reduce to $|\vec{v}_1 - \vec{v}_2|$. We leave it as an exercise to prove the result for the general case of parallel beams.

We can therefore define a machine-independent differential cross section $d\sigma$ by normalising the scattering probability by the total luminosity,

$$d\sigma = (2\pi)^4 \delta_4 \left(\sum_{i=1}^{n} p_i - \bar{k}_1 - \bar{k}_2 \right) \frac{|\mathcal{M}(\{-p_i\}, \{\bar{k}_j\})|^2}{4\sqrt{(\bar{k}_1 \cdot \bar{k}_2)^2 - m_1^2 m_2^2}} S \prod_{i=1}^{n} \frac{d_3 \vec{p}_i}{2 p_0(\vec{p}_i)(2\pi)^3}.$$

$$(10.12)$$

The parameter S is the inverse of the permutation factor for identical particles in the final state, as a detector will not be able to distinguish them. This will avoid double counting when performing the phase space integrals. If there are n_r identical particles of sort r in the final state, $S = \prod_r 1/n_r!$ (in the present case $S = 1/n!$).

Typical electromagnetic cross sections, as we will compute later, are of the order of nanobarns (1 nb= 10^{-33} cm^2). With a luminosity of 10^{33} cm^{-2}s^{-1}, approximately one collision event per second will take place. In the weak interaction the cross sections are typically five to six orders of magnitude smaller, such that not more than one event per day will take place in that case.

11

Decay Rates

DOI: 10.1201/b15364-11

The definition of the decay rate (also called decay width) of an unstable particle is best defined by considering its self-energy,

$$\Sigma(p) = \frac{1}{Z} \; \text{—}\bigcirc\!\!\!\!\!/\!\!\!\!\!\bigcirc\text{—} \, , \tag{11.1}$$

where the diagram for the $1PI$ two-point function is now to be evaluated using the Feynman rules in Table 9.1 (pg. 63). The relation with the self-energy follows from the fact that, apart from the overall factor $i/(2\pi)^4$, one has for each of the two external lines an extra factor $-i(2\pi)^2\sqrt{Z}$ as compared to the amputated $1PI$ two-point function; in total one therefore has $-iZ$ times the amputated two-point function. The latter indeed equals $i\Sigma(p)$; see Equation (9.9). We will now consider a simple example of a scalar field theory with two types of fields, a field $\varphi(x)$ associated with a light particle (mass m) and a field $\sigma(x)$ associated with a heavy particle (mass $M > 2m$), which can decay in the lighter particles if we allow for a coupling between one σ and two φ fields,

$$V(\sigma, \varphi) = \tfrac{1}{2} g \sigma \varphi^2 \, , \qquad \text{—}\!\!<^{\varphi}_{\varphi} \equiv g. \tag{11.2}$$

For the σ two-point function in lowest order we find

$$\text{—}\bigcirc\!\!\!\!\!/\!\!\!\!\!\bigcirc\text{—} = \text{—}\bigcirc\text{—} + \text{—}\ominus\text{—} + \cdots . \tag{11.3}$$

If σ is a stable particle (i.e., $M < 2m$), the loop in the first diagram corresponds to virtual φ particles moving between the vertices, since always $k^2 \neq m^2$ and $(k-p)^2 \neq m^2$. However, as soon as $M > 2m$, the loop integral will contain contributions where the φ particles can be on the mass-shell and behave as real particles, e.g., $k^2 = m^2$ and $(k-p)^2 = m^2$ in the first diagram. The real φ particles can escape to infinity, thereby describing the decay of the σ particle. Its number will reduce as a function of time.

Indeed we will see that only if $M > 2m$, the self-energy will be able to develop a nonzero imaginary part, $\gamma \equiv \text{Im}\big(-Z_\sigma \Sigma_\sigma(p)\big)\big|_{p^2=M^2} \neq 0$. On the mass-shell the σ propagator is in that case modified for $t > 0$ to

$$\int \frac{d_4 p}{(2\pi)^4} \frac{Z_\sigma e^{-ipx}}{p^2 - M^2 + i\gamma + i\varepsilon} = -\frac{i}{2} \int \frac{d_3 \vec{p}}{(2\pi)^3} \frac{Z_\sigma e^{i\vec{p}\cdot\vec{x}}}{\sqrt{\vec{p}^2 + M^2 - i\gamma}} e^{-it\sqrt{\vec{p}^2 + M^2 - i\gamma}}.$$

(11.4)

The poles $p_0 = \pm\sqrt{\vec{p}^2 + M^2 - i\gamma}$ are now complex and for $\gamma \ll M^2$ one has in a good approximation

$$e^{-it\sqrt{\vec{p}^2 + M^2 - i\gamma}} = e^{-it\sqrt{\vec{p}^2 + M^2}} e^{-\frac{1}{2}\gamma t/\sqrt{\vec{p}^2 + M^2}}.$$

(11.5)

The amplitude of the wave function for the σ particle consequently decays with a decay rate of $\Gamma(p) = \gamma/\sqrt{\vec{p}^2 + M^2}$, and the lifetime of this particle is hence $\tau(p) = 1/\Gamma(p)$.

We will evaluate the imaginary part of self-energy for the σ two-point function in Equation (11.3) first to lowest order in g

$$\text{Im}\left(-\Sigma_\sigma(p)\right) = \text{Im}\left(-\frac{ig^2}{2} \int \frac{d_4 k}{(2\pi)^4} \frac{1}{(k^2 - m^2 + i\varepsilon)((k-p)^2 - m^2 + i\varepsilon)}\right)$$

$$= \text{Im}\left(-\frac{ig^2}{2} \int d_4 x\, G^2(x) e^{ipx}\right),$$

(11.6)

where we used the definition of the Green's function, Equation (4.7). With the help of Equation (5.12) we can write

$$G(x) = -i \int \frac{d_3 \vec{k}}{(2\pi)^3} \frac{e^{i\vec{k}\cdot\vec{x}} e^{-ik_0(\vec{k})|t|}}{2k_0(\vec{k})},$$

(11.7)

which yields (after changing x to $-x$ at the right place)

$$\text{Im}\left(-\Sigma_\sigma(p)\right) = -\frac{1}{4}g^2 \int d_4 x \left(G(x)^2 + G^*(-x)^2\right) e^{ipx}$$

$$= \frac{1}{4}g^2 (2\pi)^4 \int \frac{d_3 \vec{k}_1}{2k_0^{(1)}(2\pi)^3} \frac{d_3 \vec{k}_2}{2k_0^{(2)}(2\pi)^3} \delta_3(\vec{k}_1 + \vec{k}_2 - \vec{p})$$

$$\times \left(\delta(k_0^{(1)} + k_0^{(2)} - p_0) + \delta(k_0^{(1)} + k_0^{(2)} + p_0)\right),$$

(11.8)

where we have implicitly defined $k_0^{(j)} \equiv \sqrt{\vec{k}_j^2 + m^2}$. As we wish to study the σ two-point function at $t > 0$ near the mass-shell, we can put $p_0 \equiv \sqrt{\vec{p}^2 + m^2}$ as well. In particular, restricting ourselves to $p_0 > 0$, gives

$$\Gamma(p) \equiv \frac{\text{Im}(-Z_\sigma \Sigma_\sigma(p))}{p_0} = \frac{g^2(2\pi)^4}{4p_0} \int \frac{d_3 \vec{k}_1}{2k_0^{(1)}(2\pi)^3} \frac{d_3 \vec{k}_2}{2k_0^{(2)}(2\pi)^3} \delta_4(k_1 + k_2 - p),$$

(11.9)

which is the result to lowest nontrivial order in g (to this order we can take $Z_\sigma = 1$).

To any order we can, however, decompose the part of the σ self-energy, with *two* φ particles as an intermediate state, in its $1PI$ components as follows

$$\qquad\qquad\qquad\qquad\qquad\qquad\qquad\qquad (11.10)$$

where the full φ two-point function is given by $1/[p^2 - m^2 - \Sigma_\varphi(p) + i\varepsilon]$, which on the mass-shell reduces to $Z_\varphi/(p^2 - \bar{m}^2 + i\varepsilon)$, the only part that actually contributes to $\mathrm{Im}[-Z_\sigma \Sigma_\sigma(p)]$. The $1PI$ $\varphi\varphi\sigma$ three-point function can easily be seen to be equal to $\mathcal{M}_{\varphi\varphi\sigma}(\{-k_i\}, p)/(Z_\varphi \sqrt{Z_\sigma})$ (or its complex conjugate if in- and outgoing lines are interchanged), since in lowest order it should coincide with the $\varphi\varphi\sigma$ three-point vertex. In this way we easily find the partial decay rate $d\Gamma(p)$ to be

$$d\Gamma(p) = (2\pi)^4 \delta_4(\Sigma_i k_i - p) \frac{|\mathcal{M}_{\varphi\varphi\sigma}(\{-k_i\}, p)|^2}{2p_0} S \prod_i \frac{d_3\vec{k}_i}{2k_0^{(i)}(2\pi)^3}, \qquad (11.11)$$

which, as it should be, is always positive. The symmetry factor S is the same as for the cross section in Equation (10.12). The total decay rate is found by integrating over the phase space of the outgoing particles $\Gamma(p) \equiv \int d\Gamma(p)$. The large resemblance with the formula for the cross section is no coincidence, as in both cases we have to calculate the probability for something to happen (respectively a decay or a scattering). In its present form, the formula for $\Gamma(p)$ is also valid for the decay of a particle in n other particles. The derivation is almost identical, e.g., in Equation (11.8) one now encounters $G^n(x)$ instead of $G^2(x)$ and g now stands of course for the coupling constant of n φ fields to the σ field. It is not necessary for this coupling to occur in the Lagrangian; at higher orders one can generate it from the lower couplings that do occur in the Lagrangian.

12

The Dirac Equation

DOI: 10.1201/b15364-12

To obtain a Lorentz invariant Schrödinger equation, we considered the square root of the Klein–Gordon equation. This had the disadvantage that the Hamiltonian $H = \sqrt{\vec{p}^2 + m^2}$ contains an infinite number of powers of $\vec{p}^2/m^2$, the parameter in which the square root should be expanded. It would have been better to treat space and time on a more equal footing in the Schrödinger equation. This is what Dirac took as his starting point. As the Schrödinger equation is linear in $p_0 = i\partial/\partial t$, one is looking for a Hamiltonian that is linear in the momenta $p_j = i\partial/\partial x^j (= -p^j)$.

$$i\frac{\partial \Psi}{\partial t} = H\Psi = -i\alpha_k \frac{\partial \Psi}{\partial x^k} + \beta m \Psi. \tag{12.1}$$

The question Dirac posed for himself was to find the simplest choice for α_k and β, such that the square of the Schrödinger equation gives the Klein–Gordon equation

$$p_0^2 = (-p_k\alpha_k + \beta m)^2 = \vec{p}^2 + m^2. \tag{12.2}$$

Dirac noted that only in case we allow α_k and β to be noncommuting objects (i.e., matrices) can one satisfy these equations. The above equation is equivalent to

$$\beta^2 = 1, \quad \tfrac{1}{2}(\alpha_j\alpha_k + \alpha_k\alpha_j) = 1\delta_{jk} \quad \text{and} \quad \alpha_j\beta + \beta\alpha_j = 0. \tag{12.3}$$

Historically, Dirac first considered $m \neq 0$, but the massless case ($m = 0$) is somewhat simpler, as it allows one to use $\beta = 0$ and $\alpha_k = \sigma_k$ for a solution of Equation (12.3). Here σ_k are the Pauli matrices, familiar from describing spin one-half particles.

$$\sigma_1 = \begin{pmatrix} 0 & 1 \\ 1 & 0 \end{pmatrix}, \qquad \sigma_2 = \begin{pmatrix} 0 & -i \\ i & 0 \end{pmatrix}, \qquad \sigma_3 = \begin{pmatrix} 1 & 0 \\ 0 & -1 \end{pmatrix}. \tag{12.4}$$

It is clear that two will be the smallest matrix dimension for which one can solve the equation $\tfrac{1}{2}(\alpha_j\alpha_k + \alpha_k\alpha_j) \equiv \tfrac{1}{2}\{\alpha_j, \alpha_k\} = 1\delta_{jk}$. It is not hard to prove that in a two-dimensional representation all solutions to this equation are given by

$$\alpha_j = \pm U \sigma_j U^{-1}, \tag{12.5}$$

where U is an arbitrary nonsingular complex 2×2 matrix. We should, however, require that H (and hence α_j) is Hermitian. This narrows U down to a unitary matrix, since

$$\alpha_j \alpha_j^\dagger \equiv U \sigma_j U^{-1} U^{\dagger^{-1}} \sigma_j^\dagger U^\dagger = U([U^\dagger U \sigma_j]^{-1} \sigma_j U^\dagger U) U^{-1} = 1, \tag{12.6}$$

such that $U^\dagger U \sigma_j = \sigma_j U^\dagger U$ for each j. The only 2×2 matrix that commutes with all Pauli matrices is a multiple of the identity, which proves that U is unitary (up to an irrelevant overall complex factor, which does not affect α_j).

Since the Hamiltonian is now a 2×2 matrix, the wave function $\Psi(x)$ becomes a complex two-dimensional vector, also called a *spinor*, which describes particles with spin $\hbar/2$

$$p_0 \Psi(x) = \mp \vec{p} \cdot \vec{\sigma} \, \Psi(x). \tag{12.7}$$

We have to demonstrate that the Dirac equation is covariant under Lorentz transformations. We first put the boosts to zero, because we already know from quantum mechanics how a spinor transforms under rotations

$$\vec{p} \to \vec{p}' = \exp(\vec{\omega} \cdot \vec{L})\vec{p}, \quad \Psi(x) \to \Psi'(x') = \exp\left(\frac{i}{2}\vec{\omega} \cdot \vec{\sigma}\right) \Psi(x). \tag{12.8}$$

Here L^i are real 3×3 matrices that generate the rotations in $\mathbb{R}^3$

$$L^i_{jk} = \varepsilon_{ijk}, \tag{12.9}$$

such that $[L^i, L^j] \equiv L^i L^j - L^j L^i = -\varepsilon_{ijk} L^k$. These reflect the commutation relations of the generators $i\sigma_j/2$. We will later, in the context of non-Abelian gauge theories, show that this describes the fact that SU(2) (the group of unitary transformations acting on the spinors) is a representation of SO(3) (the group of rotations in $\mathbb{R}^3$). To show the covariance of the Dirac equation under rotations, i.e.,

$$p_0 \Psi(x) = \mp \vec{p} \cdot \vec{\sigma} \, \Psi(x) \to p_0 \Psi'(x') = \mp \vec{p}' \cdot \vec{\sigma} \Psi'(x'), \tag{12.10}$$

we work out the Dirac equation in the rotated frame. Using Equation (12.8) we get

$$p_0 \Psi(x) = \mp \vec{p}' \cdot \exp\left(-\frac{i}{2}\vec{\omega} \cdot \vec{\sigma}\right) \vec{\sigma} \exp\left(\frac{i}{2}\vec{\omega} \cdot \vec{\sigma}\right) \Psi(x), \tag{12.11}$$

which should reduce to Eq. (12.7). To prove this, we use the following general result for matrices X and Y

$$e^X Y e^{-X} = \exp(\mathrm{ad}X)(Y), \quad \mathrm{ad}X(Y) \equiv [X, Y], \tag{12.12}$$

which is derived from the fact that $f_1(t) = e^{tX}Ye^{-tX}$ and $f_2(t) = \exp[\mathrm{ad}(t X)]Y$ satisfy the same differential equation, $df_i(t)/dt = [X, f_i(t)]$. Since also $f_1(0) = f_2(0)$ it follows that $f_1(1) = f_2(1)$, being the above equation. Applying this result to $Y = \sigma_k$ and $X = -i\vec{\omega} \cdot \vec{\sigma}/2$, using the fact that

$$\mathrm{ad}\left(-\frac{i}{2}\vec{\omega} \cdot \vec{\sigma}\right)\sigma_k = \left[-\frac{i}{2}\vec{\omega} \cdot \vec{\sigma}, \sigma_k\right] = (\vec{\omega} \cdot \vec{L})_{kj}\sigma_j, \tag{12.13}$$

the r.h.s. of Equation (12.11) becomes $\mp \vec{p}' \cdot \vec{\sigma}' \Psi(x) = \mp \vec{p} \cdot \vec{\sigma} \Psi(x)$, where $\vec{\sigma}' = \exp(\vec{\omega} \cdot \vec{L})\vec{\sigma}$.

The interpretation of this Schrödinger equation caused Dirac quite some trouble, as its eigenvalues are $\pm|\vec{p}|$, and it is not bounded from below. In the scalar theory we could avoid this by just considering the positive root of the Klein–Gordon equation. Only when we required localisation of the wave function inside the light cone were we forced to consider negative energy states. In the present case, restricting to one of the eigenstates would break the rotational invariance of the theory. For the massive case, Dirac first incorrectly thought that the positive energy states describe the electron and the negative energy states the proton. At that time antiparticles were unknown. Antiparticles were predicted by Dirac because the only way he could make the theory consistent was to invoke the Pauli principle and to fill all the negative energy states. A hole in this sea of negative energy states, the so-called Dirac sea, then corresponds to a state of positive energy. These holes describe the antiparticle with the same mass as the particle. Obviously the particle number will no longer be conserved and also the Dirac equation will require "second quantisation" and the introduction of a field, which will be discussed later.

For the massive Dirac equation we need to find a matrix β that anticommutes with all α_i. For 2×2 matrices this is impossible, since the Pauli matrices form a complete set of anticommuting matrices. The smallest size turns out to be a 4×4 matrix. The following representation is usually chosen

$$\alpha_i = \begin{pmatrix} \oslash & \sigma_i \\ \sigma_i & \oslash \end{pmatrix}, \quad \beta = \begin{pmatrix} 1_2 & \oslash \\ \oslash & -1_2 \end{pmatrix}, \tag{12.14}$$

for which the nonrelativistic limit has a simple form. For the massless case it is often more convenient to use the so-called Weyl representation

$$\tilde{\alpha}_i = \begin{pmatrix} \sigma_i & \oslash \\ \oslash & -\sigma_i \end{pmatrix}, \quad \tilde{\beta} = \begin{pmatrix} \oslash & -1_2 \\ -1_2 & \oslash \end{pmatrix}. \tag{12.15}$$

We leave it as an exercise for the student to show that these two representations are related by a 4×4 unitary transformation U, i.e., $\tilde{\gamma}^\mu = U\gamma^\mu U^{-1}$.

To study the covariance of the Dirac equation

$$p_0\Psi(x) = (-\alpha_i p_i + \beta m)\Psi(x) \tag{12.16}$$

under Lorentz transformations [note that now $\Psi(x)$ has four complex components], it will be profitable to introduce a 'four-vector' γ^μ of 4×4 matrices

$$\gamma^\mu \equiv (\gamma^0, \gamma^i) = (\beta, \beta\alpha_i), \tag{12.17}$$

such that the Dirac equation becomes

$$(-i\gamma^\mu \partial_\mu + m)\Psi(x) = (-\gamma^\mu p_\mu + m)\Psi(x) \equiv (-\not{p} + m)\Psi(x) = 0. \tag{12.18}$$

The Dirac gamma matrices satisfy anticommuting relations

$$\{\gamma^\mu, \gamma^\nu\} \equiv \gamma^\mu\gamma^\nu + \gamma^\nu\gamma^\mu = 2g^{\mu\nu}. \tag{12.19}$$

As for the covariance under rotations, this equation is covariant under Lorentz boosts if there exists a nonsingular complex 4×4 matrix S, such that $\Psi(x) \to \Psi'(x') = S\Psi(x)$ and

$$S^{-1}\gamma^\mu S = K^\mu{}_\nu \gamma^\nu, \tag{12.20}$$

where $K^\mu{}_\nu$ is the Lorentz transformation acting on the momenta as $p'_\mu = K_\mu{}^\nu p_\nu$ and on the coordinates as $x^{\mu'} = K^\mu{}_\nu x^\nu$. Like for the rotations, K can be written as an exponent

$$K = \exp(\omega). \tag{12.21}$$

Here $\omega^\mu{}_\nu$ is a 4×4 matrix, which is antisymmetric when one of its indices is raised or lowered by the metric $\omega_{\mu\nu} = g_{\mu\lambda}\omega^\lambda{}_\nu = -\omega_{\nu\mu}$.

We will now prove that

$$S = S(\omega) = \exp\left(-\frac{i}{4}\omega_{\mu\nu}\sigma^{\mu\nu}\right), \quad \sigma^{\mu\nu} \equiv \frac{i}{2}[\gamma^\mu, \gamma^\nu] \tag{12.22}$$

satisfies Equation (12.20). Using the antisymmetry of $\omega_{\mu\nu}$ and Equation (12.19) we find

$$\left[\frac{i}{4}\omega_{\mu\nu}\sigma^{\mu\nu}, \gamma^\lambda\right] = -\tfrac{1}{4}\omega_{\mu\nu}[\gamma^\mu\gamma^\nu, \gamma^\lambda]$$

$$= -\tfrac{1}{4}\omega_{\mu\nu}(\gamma^\mu\gamma^\nu\gamma^\lambda - \gamma^\lambda\gamma^\mu\gamma^\nu)$$

$$= \tfrac{1}{4}\omega_{\mu\nu}(\gamma^\lambda\gamma^\mu\gamma^\nu + \gamma^\mu\gamma^\lambda\gamma^\nu - 2\gamma^\mu g^{\lambda\nu})$$

$$= \tfrac{1}{2}\omega_{\mu\nu}(g^{\lambda\mu}\gamma^\nu - \gamma^\mu g^{\lambda\nu}) = \omega^\lambda{}_\mu\gamma^\mu. \tag{12.23}$$

Applying Equation (12.12) gives the proof for Equation (12.20). One says that $S(\omega)$ is a representation of $K(\omega)$. Note that in general $S(\omega)$ is not a unitary transformation. This is because the boosts form a noncompact part of the Lorentz group. There is, however, a relation between $S^\dagger$ and S^{-1},

$$\gamma^0 S^\dagger \gamma^0 = S^{-1}, \tag{12.24}$$

which is most easily proven in the Weyl representation, since $\tilde{\gamma}^0 = \tilde{\beta}$ commutes with $\tilde{\sigma}_{ij}$ but *anticommutes* with $\tilde{\sigma}_{k0}$ (as always Roman indices run from

1 to 3 and Greek indices run from 0 to 3), whereas $\tilde{\sigma}_{ij}$ is Hermitian and $\tilde{\sigma}_{k0}$ is *anti*-Hermitian, as follows from the explicit expressions obtained from Equations (12.15), (12.17), and (12.22)

$$\tilde{\sigma}_{ij} = \tilde{\sigma}^{ij} = \frac{i}{2}[\tilde{\gamma}^i, \tilde{\gamma}^j] = \varepsilon_{ijk} \begin{pmatrix} \sigma_k & \oslash \\ \oslash & \sigma_k \end{pmatrix},$$

$$\tilde{\sigma}_{0k} = \tilde{\sigma}^{k0} = \frac{i}{2}[\tilde{\gamma}^k, \tilde{\gamma}^0] = -i \begin{pmatrix} \sigma_k & \oslash \\ \oslash & -\sigma_k \end{pmatrix}. \tag{12.25}$$

Note that Equations (12.20) and (12.22) are independent of the representation in which we give the gamma matrices, as any two such representations have to be related by a unitary transformation. In the Weyl representation $\tilde{S}(\omega)$ is block diagonal, like the Dirac equation for $m = 0$ (as $\tilde{\alpha}_i$ is block diagonal). The upper block corresponds to Equation (12.7) with the plus sign and the lower block corresponds to the minus sign. We can verify Equation (12.8) by using the fact that the Lorentz transformations contain the rotations through the identification $\omega_k = -\frac{1}{2}\varepsilon_{ijk}\omega_{ij}$. With $\omega_{0k} = 0$, one finds $\tilde{S}(\omega) = 1_2 \otimes \exp(i\vec{\omega}\cdot\vec{\sigma}/2)$, i.e., it acts on each 2×2 block by the same unitary transformation.

The boost parameters are described by ω_{0i}. For a boost in the x direction, we have that $\chi = \omega_{01}$ is related to the boost velocity by $v_1 = -\tanh(\chi)$. For K we find in this case

$$K = \begin{pmatrix} \cosh(\chi) & \sinh(\chi) & 0 & 0 \\ \sinh(\chi) & \cosh(\chi) & 0 & 0 \\ 0 & 0 & 1 & 0 \\ 0 & 0 & 0 & 1 \end{pmatrix}. \tag{12.26}$$

In the Weyl representation, $\tilde{S}$ splits again in two blocks, but one is the inverse of the other (and neither is unitary). To be precise, $\tilde{S}$ restricted to the upper-left 2×2 block equals $\exp(\frac{1}{2}\omega_{0k}\sigma_k)$, whereas for the lower-right 2×2 block we find $\exp(-\frac{1}{2}\omega_{0k}\sigma_k)$.

As S is not unitary, $\Psi^\dagger(x)\Psi(x)$ is no longer invariant under Lorentz transformations. But we claim it is nevertheless a probability density, namely the time component $j^0(x)$ of a conserved current

$$j^\mu(x) = \Psi^\dagger(x)\gamma^0\gamma^\mu\Psi(x). \tag{12.27}$$

We leave it as an exercise to show that the Dirac equation implies that the current is conserved, $\partial_\mu j^\mu(x) = 0$. The combination $\Psi^\dagger(x)\gamma^0$ will occur so often that it has acquired its own symbol

$$\overline{\Psi}(x) \equiv \Psi^\dagger(x)\gamma^0. \tag{12.28}$$

It transforms under a Lorentz transformation as

$$\Psi(x) \rightarrow S\Psi(x) \quad \text{and} \quad \overline{\Psi}(x) \rightarrow \overline{\Psi}(x)S^{-1}. \tag{12.29}$$

We can use this to build the required Lorentz scalars, vectors, and tensors

scalar: $\quad \overline{\Psi}'(x')\Psi'(x') = \overline{\Psi}(x)S^{-1}S\Psi(x) \quad = \overline{\Psi}(x)\Psi(x)$

vector: $\quad \overline{\Psi}'(x')\gamma^\mu\Psi'(x') = \overline{\Psi}(x)S^{-1}\gamma^\mu S\Psi(x) = K^\mu{}_\nu \overline{\Psi}(x)\gamma^\nu\Psi(x)$

tensor: $\quad \overline{\Psi}'(x')\sigma^{\mu\nu}\Psi'(x') = \overline{\Psi}(x)S^{-1}\sigma^{\mu\nu}S\Psi(x) = K^\mu{}_\lambda K^\nu{}_\kappa \overline{\Psi}(x)\sigma^{\lambda\kappa}\Psi(x).$

The Lagrangian is a Lorentz scalar, which we chose such that its equations of motion reproduce the Dirac equation. As $\Psi(x)$ is complex, it can be considered independent of $\overline{\Psi}(x)$ and the following Lagrangian:

$$S_{\text{Dirac}} = \int d_4x \, \mathcal{L}_{\text{Dirac}} = \int d_4x \, \overline{\Psi}(x)(i\gamma^\mu\partial_\mu - m)\Psi(x). \tag{12.30}$$

gives the Euler–Lagrange equations

$$\frac{\delta S}{\delta \overline{\Psi}(x)} = (i\gamma^\mu\partial_\mu - m)\Psi(x) = 0, \quad \frac{\delta S}{\delta \Psi(x)} = \overline{\Psi}(x)(-i\gamma^\mu\overleftarrow{\partial}_\mu - m) = 0. \tag{12.31}$$

The second equation is the complex conjugate of the first, $\Psi^\dagger(x)[i(\gamma^\mu)^\dagger\overleftarrow{\partial}_\mu + m] = 0$, because the gamma matrices satisfy

$$(\gamma^\mu)^\dagger = \gamma^0\gamma^\mu\gamma^0, \tag{12.32}$$

which follows from the fact that $(\gamma^0)^\dagger = \beta^\dagger = \beta = \gamma^0$ and $(\gamma^i)^\dagger = (\beta\alpha_i)^\dagger = \alpha_i\beta = -\gamma^i$, or from the explicit representation of the gamma matrices

$$\gamma^0 = \begin{pmatrix} 1_2 & \oslash \\ \oslash & -1_2 \end{pmatrix}, \quad \gamma^i = \begin{pmatrix} \oslash & \sigma_i \\ -\sigma_i & \oslash \end{pmatrix}. \tag{12.33}$$

Hence

$$0 = \Psi^\dagger(x)(i(\gamma^\mu)^\dagger\overleftarrow{\partial}_\mu + m)\gamma^0 = \overline{\Psi}(i\gamma^0(\gamma^\mu)^\dagger\gamma^0\overleftarrow{\partial}_\mu + m)$$

$$= \overline{\Psi}(i\gamma^\mu\overleftarrow{\partial}_\mu + m). \tag{12.34}$$

An important role will be played by a fifth gamma matrix

$$\gamma_5 \equiv i\gamma^0\gamma^1\gamma^2\gamma^3 = \begin{pmatrix} \oslash & 1_2 \\ 1_2 & \oslash \end{pmatrix}, \tag{12.35}$$

which anticommutes with all γ^μ (see Problem 21)

$$\gamma_5\gamma^\mu = -\gamma^\mu\gamma_5 \quad \text{and} \quad (\gamma_5)^2 = 1_4. \tag{12.36}$$

This implies that we can introduce projection operators

$$P_\pm \equiv \tfrac{1}{2}(1 \pm \gamma_5), \tag{12.37}$$

which satisfy $(P_\pm)^2 = P_\pm$ and $P_\pm P_\mp = 0$. Their role is best described in the Weyl representation, where

$$\tilde{\gamma}_5 = i\tilde{\gamma}_0\tilde{\gamma}_1\tilde{\gamma}_2\tilde{\gamma}_3 = -i\tilde{\alpha}_1\tilde{\alpha}_2\tilde{\alpha}_3 = \begin{pmatrix} 1_2 & \oslash \\ \oslash & -1_2 \end{pmatrix}, \tag{12.38}$$

such that $\tilde{P}_+ = \tfrac{1}{2}(1 + \tilde{\gamma}_5)$ projects on the two upper and $\tilde{P}_- = \tfrac{1}{2}(1 - \tilde{\gamma}_5)$ on the two lower components of the four spinors. In the massless case these two components are decoupled,

$$p_0 \Psi(x) = (-\vec{p} \cdot \vec{\tilde{\alpha}} + m\tilde{\beta})\Psi(x) = \begin{pmatrix} -\vec{p} \cdot \vec{\sigma} & -m1_2 \\ -m1_2 & \vec{p} \cdot \vec{\sigma} \end{pmatrix} \Psi. \tag{12.39}$$

Hence, for $m = 0$ we have

$$\Psi(x) = \begin{pmatrix} \Psi_+(x) \\ \Psi_-(x) \end{pmatrix}, \qquad p_0 \Psi_\pm(x) = \mp \vec{p} \cdot \vec{\sigma} \Psi_\pm(x). \tag{12.40}$$

which is identical to Equation (12.7). The eigenstates of the projection operators $P_\pm$ are called helicity eigenstates. As long as $m \neq 0$, helicity is not conserved. But as we saw, for $m = 0$ the two helicity eigenstates decouple. One can define in that case consistently a particle with a fixed helicity, whose opposite helicity state does not occur (although its antiparticle has opposite helicity). A very important example of such a particle is the neutrino, although experiment has not yet been able to rule out a (tiny) mass for this particle ($m_{\nu_e} < 10eV$). See Problem 22 for more details.

Apart from the invariance of the Dirac equation under Lorentz transformations and translations (which are obvious symmetries of $\mathcal{L}_{\text{Dirac}}$), we also often want invariance under parity ($\vec{x} \to -\vec{x}$) and time reversal ($t \to -t$). One easily checks that

$$P\Psi(x) \equiv \Psi'(t, -\vec{x}) = \gamma_0 \Psi(t, \vec{x}) \quad \text{and} \quad T\Psi(x) \equiv \Psi'(-t, \vec{x})$$
$$= \gamma_5 \gamma_0 \gamma^2 \Psi^*(t, \vec{x}) \tag{12.41}$$

satisfy the Dirac equation, where P stands for parity and T for time reversal. This implies that the Lorentz covariant combinations $\overline{\Psi}(x)\gamma_5\Psi(x)$ and $\overline{\Psi}(x)\gamma_5\gamma^\mu\Psi(x)$ are not invariant under parity and time reversal. They are called pseudoscalars and pseudovectors. These combinations play an important role in the weak interactions, where parity is not a symmetry. A third discrete symmetry, charge conjugation C, will be discussed in Chapter 17.

13

Plane Wave Solutions of the Dirac Equation

DOI: 10.1201/b15364-13

As usual, the Dirac equation can be solved by Fourier decomposition in plane waves,

$$\Psi(x) = \int \frac{d_3\vec{k}}{\sqrt{2k_0(\vec{k})(2\pi)^3}} \, \tilde{\Psi}(\vec{k})e^{-ikx}. \tag{13.1}$$

where $\tilde{\Psi}(\vec{k})$ are complex four-vectors that satisfy

$$(\not{k} - m)\tilde{\Psi}(\vec{k}) \equiv (k_\mu \gamma^\mu - m)\tilde{\Psi}(\vec{k}) = 0, \qquad k_0^2 = \vec{k}^2 + m^2. \tag{13.2}$$

The Lorentz invariance implies

$$\tilde{\Psi}'(\vec{k}') = S(\omega)\tilde{\Psi}(\vec{k}), \qquad k'_\mu = K(\omega)_\mu{}^\nu k_\nu. \tag{13.3}$$

As any $\vec{k}$ can be obtained from a boost to $\vec{k} = \vec{0}$, all solutions of the Dirac equation can be obtained from the ones at rest with $\vec{k} = \vec{0}$ (see Problem 19)

$$(\gamma^0 k_0 - m)\tilde{\Psi}(\vec{0}) = \begin{pmatrix} (k_0 - m)1_2 & \oslash \\ \oslash & -(k_0 + m)1_2 \end{pmatrix} \tilde{\Psi}(\vec{0}). \tag{13.4}$$

We see that for $k_0 > 0$ ($k_0 = m$), there are two independent solutions both of the form

$$\tilde{\Psi}_+(\vec{0}) = \begin{pmatrix} \Psi_A \\ 0 \end{pmatrix}, \tag{13.5}$$

where Ψ_A is a spin one-half two-spinor of which $\begin{pmatrix} 1 \\ 0 \end{pmatrix}$ is the spin-up and $\begin{pmatrix} 0 \\ 1 \end{pmatrix}$ is the spin-down state. The identification of the spin degrees of freedom follows from the behaviour of $\tilde{\Psi}$ under rotations (which leave $\vec{k} = \vec{0}$). We leave it as an exercise to verify that also in the Dirac representation of the gamma matrices, like in the Weyl representation [see Equation (12.25)],

$$\sigma_{ij} = \varepsilon_{ijk} \begin{pmatrix} \sigma_k & \oslash \\ \oslash & \sigma_k \end{pmatrix},$$
(13.6)

such that Ψ_A is easily seen to transform under a rotation as in Eq. (12.8) [compare this to the discussion following Equation (12.25)]. There are also two solutions for $k_0 < 0$ ($k_0 = -m$) of the form

$$\tilde{\Psi}_-(\vec{0}) = \begin{pmatrix} 0 \\ \Psi_B \end{pmatrix},$$
(13.7)

where likewise Ψ_B is a spin one-half two-spinor of which $\begin{pmatrix} 1 \\ 0 \end{pmatrix}$ is the spin-up

and $\begin{pmatrix} 0 \\ 1 \end{pmatrix}$ is the spin-down state, which transform under rotations as Ψ_A.

For any frame, i.e., for any value of $\vec{k}$, we will define the four independent solutions of the Dirac equation as

$$k_0 = \sqrt{\vec{k}^2 + m^2} : \quad u^{(\alpha)}(\vec{k}) = \frac{\not{k} + m}{\sqrt{m + |k_0|}} u_0^{(\alpha)}, \quad u_0^{(1)} = \begin{pmatrix} 1 \\ 0 \\ 0 \\ 0 \end{pmatrix}, \quad u_0^{(2)} = \begin{pmatrix} 0 \\ 1 \\ 0 \\ 0 \end{pmatrix},$$

$$k_0 = -\sqrt{\vec{k}^2 + m^2} : \quad v^{(\alpha)}(-\vec{k}) = \frac{\not{k} + m}{\sqrt{m + |k_0|}} v_0^{(\alpha)}, \quad v_0^{(1)} = \begin{pmatrix} 0 \\ 0 \\ 1 \\ 0 \end{pmatrix}, \quad v_0^{(2)} = \begin{pmatrix} 0 \\ 0 \\ 0 \\ 1 \end{pmatrix}.$$
(13.8)

These solutions naturally split in positive energy [$u^{(\alpha)}(\vec{k})$ with $\alpha = 1, 2$] and negative energy [$v^{(\alpha)}(-\vec{k})$ with $\alpha = 1, 2$] solutions; for $\vec{k} = \vec{0}$ and $m \neq 0$ easily seen to be proportional to the solutions $\tilde{\Psi}_\pm(\vec{0})$, to be precise $u^{(\alpha)}(\vec{0}) = u_0^{(\alpha)}/\sqrt{2m}$ and $v^{(\alpha)}(\vec{0}) = v_0^{(\alpha)}/\sqrt{2m}$. The normalisation we have chosen allows us to treat massless fermions at the same footing. In that case we can, however, not transform to the rest frame. This normalisation also implies that

$$\overline{u^{(\alpha)}(\vec{k})} u^{(\beta)}(\vec{k}) = -\overline{v^{(\alpha)}(-\vec{k})} v^{(\beta)}(-\vec{k}) = 2m\delta_{\alpha\beta}.$$
(13.9)

For example,

$$\overline{u^{(\alpha)}(\vec{k})}u^{(\beta)}(\vec{k}) = \frac{\overline{u_0^{(\alpha)}}\gamma^0(\not{k}+m)^\dagger\gamma^0(\not{k}+m)u_0^{(\beta)}}{(m+k_0)} = \frac{\overline{u_0^{(\alpha)}}(\not{k}+m)^2u_0^{(\beta)}}{(m+k_0)}$$

$$= \frac{\overline{u_0^{(\alpha)}}(k^2+m^2+2\not{k}m)u_0^{(\beta)}}{(m+k_0)} = \frac{\overline{u_0^{(\alpha)}}(k^2+m^2+2k_0m)u_0^{(\beta)}}{(m+k_0)}$$

$$= 2m\delta_{\alpha\beta}, \tag{13.10}$$

where we used $k^2 = m^2$ and $\overline{u_0^{(\alpha)}}\gamma^\mu u_0^{(\beta)} = \delta_{\alpha\beta}\delta_{\mu0}$ [see Equation (12.33)]. The computation for $v^{(\alpha)}(\vec{k})$ is left as an exercise. The fact that we find a result that is independent of $\vec{k}$ is consistent with our claim that these spinors can also be obtained by applying the appropriate boost to $\vec{k} = \vec{0}$, since $\overline{\Psi}(\vec{k})\Psi(\vec{k})$ is a Lorentz scalar.

That these spinors indeed satisfy Equation (13.2) follows from the fact that

$$(\not{k}-m)(\not{k}+m) = k^2 - m^2 = 0. \tag{13.11}$$

Note that $v^{(\alpha)}(\vec{k})$ can also be viewed as a positive energy solution for the complex conjugate of the Dirac equation

$$k_0 = \sqrt{\vec{k}^2+m^2}: \quad (\not{k}+m)v^{(\alpha)}(\vec{k}) = 0. \tag{13.12}$$

It will play the role of the wave function for the antiparticles (the holes) in the Dirac field theory. To see that our plane waves have the correct amplitude, we use the fact that the probability density can be defined in terms of the zero-component $\rho(x)$ of the conserved four-vector $j^\mu(x) = \overline{\Psi}(x)\gamma^\mu\Psi(x)$ [see Equation (12.27)], or

$$\rho(x) = \Psi^\dagger(x)\Psi(x). \tag{13.13}$$

Indeed, $\Psi^\dagger(\vec{k})\Psi(\vec{k})$ transforms as a density, i.e., as the energy k_0, and we find

$$u^{(\alpha)}(\vec{k})^\dagger u^{(\beta)}(\vec{k}) = v^{(\alpha)}(\vec{k})^\dagger v^{(\beta)}(\vec{k}) = 2k_0\delta_{\alpha\beta}, \quad u^{(\alpha)}(\vec{k})^\dagger v^{(\beta)}(\vec{k}) = 0. \tag{13.14}$$

This can be verified by a direct computation, e.g., $(k_0 > 0)$,

$$v^{(\alpha)}(\vec{k})^\dagger v^{(\beta)}(\vec{k}) = \frac{v_0^{(\alpha)\dagger}(m-\not{k})^\dagger(m-\not{k})v_0^{(\beta)}}{m+k_0} = \frac{v_0^{(\alpha)\dagger}(k_0^2+\vec{k}^2+m^2-2mk_0\gamma^0)v_0^{(\beta)}}{m+k_0}$$

$$= \frac{v_0^{(\alpha)\dagger}(k_0^2+\vec{k}^2+m^2+2mk_0)v_0^{(\beta)}}{m+k_0} = 2k_0\delta_{\alpha\beta}, \tag{13.15}$$

using $v_0^{(\alpha)\dagger}\gamma^\mu v_0^{(\beta)} = -\delta_{\mu0}\delta_{\alpha\beta}$ [see Equation (12.33)]. We leave the other identities as an exercise.

This implies that the plane wave solutions $(k_0 = (\vec{k}^2 + m^2)^{\frac{1}{2}})$

$$\frac{u^{(\alpha)}(\vec{k})}{\sqrt{2k_0(2\pi)^3}}e^{-ikx}, \qquad \frac{v^{(\alpha)}(\vec{k})}{\sqrt{2k_0(2\pi)^3}}e^{ikx}, \qquad \alpha = 1,\ 2, \qquad (13.16)$$

are normalised to one [of course, in a finite volume we replace $(2\pi)^3$ by V and $\int d_3\vec{k}$ by $\sum_{\vec{k}}$]. As for the scalar field, we can introduce a Dirac field

$$\Psi(x) = \int \frac{d_3\vec{k}}{\sqrt{2k_0(\vec{k})(2\pi)^3}} \sum_{\alpha=1}^{2} \left(b_\alpha(\vec{k})u^{(\alpha)}(\vec{k})e^{-ikx} + d_\alpha^\dagger(\vec{k})v^{(\alpha)}(\vec{k})e^{ikx} \right). \quad (13.17)$$

Since $\Psi(x)$ is complex, there is no relation between d_α and b_α. In the quantum field theory, these will play the role of the annihilation operators for the (anti-)particles.

Finally we note that, for $m \neq 0$, we can define projection operators (which are 4×4 matrices)

$$\Lambda_+(\vec{k}) = \sum_{\alpha=1}^{2} \frac{u^{(\alpha)}(\vec{k}) \otimes \overline{u^{(\alpha)}(\vec{k})}}{2m} = \frac{\slashed{k}+m}{2m},$$

$$\Lambda_-(\vec{k}) = -\sum_{\alpha=1}^{2} \frac{v^{(\alpha)}(\vec{k}) \otimes \overline{v^{(\alpha)}(\vec{k})}}{2m} = \frac{-\slashed{k}+m}{2m}, \qquad (13.18)$$

which can be verified from the explicit form of u and v, or by first computing them in the rest frame where $\Lambda_\pm(\vec{0}) = \frac{1}{2}(1 \pm \gamma^0)$. In that case we find

$$\Lambda_\pm(\vec{k}) = \frac{(\pm\slashed{k}+m)(1\pm\gamma^0)(\pm\slashed{k}+m)}{4m(m+k_0)} = \frac{\pm\slashed{k}+m}{2m}, \qquad (13.19)$$

whose proof requires some gamma matrix gymnastics. Independent of the frame $\Lambda_\pm$ satisfy

$$\Lambda_\pm^2(\vec{k}) = \Lambda_\pm(\vec{k}), \qquad \Lambda_\mp(\vec{k})\Lambda_\pm(\vec{k}) = 0. \qquad (13.20)$$

Note that $\text{Tr}(\Lambda_\pm(\vec{k})) = 2$, a result that can also be derived from Equation (13.9).

14

The Dirac Hamiltonian

DOI: 10.1201/b15364-14

In the Dirac equation we encountered an additional difficulty, namely that the negative energy solutions even arise at the level of the classical theory. In field theory the negative energy solutions had an interpretation in terms of antiparticles, and the field theory Hamiltonian was still positive, and most importantly, bounded from below (see Chapter 2 and Problem 5). The field theory Hamiltonian for the Dirac field no longer has this property. The Hamiltonian can again be derived through a Legendre transform of the Lagrangian

$$S = \int d_4x\, \mathcal{L} = \int d_4x\, \overline{\Psi}(x)(i\gamma^\mu \partial_\mu - m)\Psi(x). \tag{14.1}$$

The canonical momentum is hence

$$\pi_a(x) = \frac{\delta S}{\delta \dot{\Psi}_a(x)} = \left(\overline{\Psi}(x) i\gamma^0\right)_a = i\Psi_a^*(x), \tag{14.2}$$

such that

$$
\begin{aligned}
H &= \int d_3\vec{x}\, \left(\pi_a(x)\dot{\Psi}_a(x) - \mathcal{L}\right) = \int d_3\vec{x}\, \overline{\Psi}(x)(-i\gamma^j \partial_j + m)\Psi(x) \\
&= \int d_3\vec{x}\, \Psi^\dagger(x)(-i\alpha_j \partial_j + m\beta)\Psi(x) \\
&= \int d_3\vec{k}\, k_0(\vec{k}) \sum_\alpha \left(b_\alpha^\dagger(\vec{k})b_\alpha(\vec{k}) - d_\alpha(\vec{k})d_\alpha^\dagger(\vec{k})\right).
\end{aligned}
\tag{14.3}
$$

Note the resemblance with Equation (12.1) for the middle term. We used Equation (13.17) for the expansion of the Dirac field in plane waves. From this result it is clear that the Hamiltonian is not bounded from below, and this would make the vacuum unstable, as the negative energy states, described by the $d_\alpha(\vec{k})$, can lower the energy by an arbitrary amount. It is well known how Dirac repaired this problem. He postulated that all negative energy states are occupied, and that the states satisfy the Pauli principle, i.e., two particles cannot occupy the same quantum state. (It is only in that case that we can make sense of what is meant with filling all negative energy states.) This implies that one should use anticommuting relations for the creation and annihilation operators

$$\{b_\alpha(\vec{k}), b_\beta(\vec{p})\} = 0, \{b_\alpha(\vec{k}), b_\beta^\dagger(\vec{p})\} = \delta_{\alpha\beta}\delta_3(\vec{k} - \vec{p}),$$

$$\{d_\alpha(\vec{k}), d_\beta(\vec{p})\} = 0, \{d_\alpha(\vec{k}), d_\beta^\dagger(\vec{p})\} = \delta_{\alpha\beta}\delta_3(\vec{k} - \vec{p}),$$

$$\{d_\alpha(\vec{k}), b_\beta(\vec{p})\} = 0, \{d_\alpha(\vec{k}), b_\beta^\dagger(\vec{p})\} = 0. \qquad (14.4)$$

Indeed, if we define a two-particle state as $|\vec{k}, \vec{p}>\equiv b^\dagger(\vec{k})b^\dagger(\vec{p})|0>$ (suppressing the spinor indices), the anticommutation relations imply that $|\vec{k}, \vec{p}>=-|\vec{p}, \vec{k}>$.

A hole in the Dirac sea is by definition the state that is obtained by annihilating a negative energy state in the Dirac sea. As annihilation lowers the total energy by the energy of the annihilated state, which in this case is negative, the net energy is raised. The wave function for the negative energy state is given by $\exp(ikx)v^{(\alpha)}(\vec{k})/\sqrt{k_0}$ and has momentum $-\vec{k}$ $[k_0 \equiv (\vec{k}^2 + m^2)^{\frac{1}{2}}]$; see Equation (13.8). The reason to associate its Fourier coefficient in Equation (13.17) with a creation operator $d_\alpha^\dagger(\vec{k})$, is that conservation of energy and momentum implies that it creates an antiparticle as a hole in the Dirac sea, with momentum $\vec{k}$ and helicity $\frac{1}{2}$ for $\alpha = 2$, whereas for $\alpha = 1$ the helicity is $-\frac{1}{2}$ (hence the helicity and momentum are opposite to the negative energy state it annihilates). The wave function of an antiparticle with momentum $\vec{k}$ is hence given by $\exp(-ikx)v^{(\alpha)}(\vec{k})^\dagger/\sqrt{k_0}$. If we now use the anticommutation relations of creation and annihilation operators (as a consequence of the Pauli exclusion principle), we see that with the present interpretation the Dirac Hamiltonian is bounded from below

$$H = \int d_3\vec{k} \, k_0(\vec{k}) \sum_\alpha \left(b_\alpha^\dagger(\vec{k})b_\alpha(\vec{k}) + d_\alpha^\dagger(\vec{k})d_\alpha(\vec{k}) - 1 \right). \qquad (14.5)$$

As in the case for scalar field theory, we can normalise the energy of the vacuum state to zero by adding a (infinite) constant. Note that this constant has its sign opposite to the scalar case (and is in magnitude *four* times as large). In so-called supersymmetric field theories, this is no longer an accident as the Dirac fields will be related to the scalar fields by a symmetry, which is, however, outside the scope of these lectures.

It is also important to note that the anticommuting relations are crucial to guarantee locality of the Dirac field. In this case, Dirac fields specified in different regions of space-time that are space-like separated should anticommute. And indeed, in Problem 24 you are asked to prove that

$$\{\Psi_a(x), \Psi_b^\dagger(x')\} = 0 \quad \text{for} \quad (x - x')^2 < 0. \qquad (14.6)$$

Also, as in the scalar theory, we can couple a source to the free Dirac field (we again introduce $\bar{\varepsilon}$ as the expansion parameter for taking the interactions due to the source into account in Hamiltonian perturbation theory)

$$\mathcal{L}(x) = \overline{\Psi}(x)(i\gamma^\mu\partial_\mu - m)\Psi(x) - \bar{\varepsilon}\bar{\mathcal{J}}(x)\Psi(x) - \bar{\varepsilon}\overline{\Psi}(x)\mathcal{J}(x),$$

$$\mathcal{H}(x) = \overline{\Psi}(x)(-i\gamma^k\partial_k + m)\Psi(x) + \bar{\varepsilon}\bar{\mathcal{J}}(x)\Psi(x) + \bar{\varepsilon}\overline{\Psi}(x)\mathcal{J}(x), \quad (14.7)$$

where, as before, the Hamiltonian H is the spatial integral over the Hamiltonian density $\mathcal{H}(x)$, $H = \int d_3\vec{x}\, \mathcal{H}(x)$. Note that the sources $\mathcal{J}(x)$ and $\bar{\mathcal{J}}(x)$ are independent, as for a complex scalar field, see Problem 17. In Problem 24 you are asked to prove that in Hamiltonian perturbation theory one obtains

$$< 0|T\exp\left(-i\int_0^T H(t)dt\right)|0 > e^{iE_0T} = 1 - i\bar{\varepsilon}^2\int d_4x d_4y\, \bar{\mathcal{J}}(x)G_F(x-y)$$

$$\times \mathcal{J}(y) + \mathcal{O}(\bar{\varepsilon}^3)$$

$$= 1 - i\; \underset{\bar{\varepsilon}\mathcal{J}}{\times}\!\!\longrightarrow\!\!\underset{\bar{\varepsilon}\bar{\mathcal{J}}}{\times}\; + \mathcal{O}(\bar{\varepsilon}^3)$$

$$(14.8)$$

where

$$G_F(x-y) = \int \frac{d_4p}{(2\pi)^4}\, \frac{e^{-ip(x-y)}}{\not{p} - m + i\varepsilon} = \int \frac{d_4p}{(2\pi)^4}\, \frac{(\not{p}+m)e^{-ip(x-y)}}{p^2 - m^2 + i\varepsilon}. \quad (14.9)$$

Hence, the Green's function for fermions is in Fourier space given by $G_F(p) = (\not{p} - m + i\varepsilon)^{-1}$, which is the inverse of the quadratic part of the Lagrangian, as for the scalar fields. In Problem 24 it will be evident that, nevertheless, the anticommuting properties of the Dirac field play a crucial role (compare this to chapter 5). It becomes, however, plausible that there is *also* for the fermions a path integral formulation, as splitting of a square [compare this to Equations (7.11) and (7.12)] is independent of the details of the path integrals. This will be the subject of the next chapter.

To conclude, we note that the coupling to an electromagnetic field $A_\mu(x)$ should be achieved through the current $j^\mu(x)$, defined in Equation (12.27)

$$\int d_4x\, j^\mu(x)A_\mu(x) = \int d_4x\, \overline{\Psi}(x)\gamma^\mu A_\mu(x)\Psi(x). \quad (14.10)$$

Since this current is conserved, which can also be seen as a consequence of the Noether theorem applied to the invariance of Equation (14.1) (the free Dirac Lagrangian) under global phase transformations, the coupling is gauge invariant. Using the minimal coupling defined as in Equation (3.35),

$$D_\mu\Psi(x) \equiv (\partial_\mu - ieA_\mu(x))\Psi(x), \quad (14.11)$$

one can fix the normalisation of the electromagnetic current to be $J^\mu(x) = -ej^\mu(x)$, since [compare this to Equation (3.31)]

$$\overline{\Psi}(x)(i\gamma^\mu D_\mu - m)\Psi(x) = \overline{\Psi}(x)(i\gamma^\mu\partial_\mu - m)\Psi(x) + ej^\mu A_\mu(x). \quad (14.12)$$

The current $j^\mu(x)$ is the charge density current, whose time component $\rho(x)$ at the quantum level is no longer positive definite. Using the anticommuting properties of Equation (14.4) one finds

$$
\begin{aligned}
\int d_3\vec{x}\ \hat{\rho}(\vec{x}) &= \int d_3\vec{x}\ \Psi^\dagger(x)\Psi(x) \\
&= \int d_3\vec{k}\ \sum_\alpha \left(b_\alpha^\dagger(\vec{k})b_\alpha(\vec{k}) + d_\alpha(\vec{k})d_\alpha^\dagger(\vec{k}) \right) \\
&= \int d_3\vec{k}\ \sum_\alpha \left(b_\alpha^\dagger(\vec{k})b_\alpha(\vec{k}) - d_\alpha^\dagger(\vec{k})d_\alpha(\vec{k}) + 1 \right) \\
&= Q_0/e + \int d_3\vec{k}\ \sum_\alpha \left(b_\alpha^\dagger(\vec{k})b_\alpha(\vec{k}) - d_\alpha^\dagger(\vec{k})d_\alpha(\vec{k}) \right). \quad (14.13)
\end{aligned}
$$

The vacuum value of this operator is indicated by the (generally infinite) constant Q_0/e, which can be normalised to zero. After all, we want the state with all negative energy states occupied to have zero charge. With the above normalisation of the electric current, $J^\mu(x) = -ej^\mu(x)$, we see that the b modes can be identified with the electrons with charge $-e$ and the d modes with their antiparticles, the positrons, with opposite electric charge $+e$. To summarise, $b_\alpha^\dagger(\vec{k})$ corresponds with the creation operator of a spin-up ($\alpha = 1$) or a spin-down ($\alpha = 2$) electron of momentum $\vec{k}$, whereas $d_\alpha^\dagger(\vec{k})$ corresponds with the creation operator of a spin-up ($\alpha = 2$) or a spin-down ($\alpha = 1$) positron of momentum $\vec{k}$.

15

Path Integrals for Fermions

DOI: 10.1201/b15364-15

For scalar fields, which describe bosons, we used real or complex numbers (the eigenvalues of the operators) in order to perform the path integral. For fermionic fields it is essential to build the anticommuting properties into the path integral.

To this end we introduce a so-called Grassmann algebra, which exists of Grassmann variables θ_i that mutually anticommute

$$\{\theta_i, \theta_j\} \equiv \theta_i\theta_j + \theta_j\theta_i = 0. \tag{15.1}$$

In particular, a Grassmann variable squares to zero

$$\theta^2 = 0. \tag{15.2}$$

A Grassmann variable can be multiplied by a complex number, with which it commutes. A function of a single Grassmann variable has a finite Taylor series

$$f(\theta) = a_0 + a_1\theta, \tag{15.3}$$

and spans a two-dimensional (real or) complex vector space. This is exactly what we need to describe a spin one-half particle. Let us introduce the following notation:

$$|0> = \begin{pmatrix} 1 \\ 0 \end{pmatrix}, \qquad |1> = \begin{pmatrix} 0 \\ 1 \end{pmatrix}. \tag{15.4}$$

With respect to the Hamiltonian

$$H_0 = \tfrac{1}{2} \begin{pmatrix} -W & 0 \\ 0 & W \end{pmatrix}, \tag{15.5}$$

$|0>$ is the vacuum state (i.e., the state with lowest energy) and we interpret $|1>$ as the one-particle state (with energy W above the vacuum). An arbitrary spinor can be written as a linear combination of these two states

$$|\Psi> = a_0|0> + a_1|1> = a_0|0> + a_1 b^\dagger|0>, \tag{15.6}$$

where b is the fermionic annihilation operator, which in the spinor representation is given by a 2×2 matrix

$$b = \begin{pmatrix} 0 & 1 \\ 0 & 0 \end{pmatrix}, \qquad b^\dagger = \begin{pmatrix} 0 & 0 \\ 1 & 0 \end{pmatrix}. \tag{15.7}$$

We note that $b^2 = (b^\dagger)^2 = 0$, a property it has in common with a Grassmann variable. We will now look for properties of θ such that

$$\Psi(\theta) \equiv < \theta | \Psi > = a_0 + a_1 \theta \tag{15.8}$$

is a representation of the state $| \Psi >$, similar to $\Psi(x) = < x | \Psi >$ for the case of a single bosonic (i.e., commuting, as opposed to anticommuting) degree of freedom. In the latter case, the normalisation

$$\int dx \ \Psi^*(x) \Psi(x) = 1 \tag{15.9}$$

is an important property we would like to impose here too (keeping in mind that for path integrals we need to insert completeness relations). As $< 0| = (1, 0)$ and $< 1| = (0, 1)$, we have

$$< \Psi | = < 0 | a_0^* + < 1 | a_1^* = < 1 | b^\dagger a_0^* + < 1 | a_1^*. \tag{15.10}$$

As in Equation (15.8) (i.e., $b^\dagger \to \theta$), we anticipate

$$\Psi^*(\theta) \equiv < \Psi | \theta > = a_0^* \theta + a_1^*. \tag{15.11}$$

We wish to define integration over Grassmann variables, such that the normalisation of the wave function is as usual

$$< \Psi | \Psi > = |a_0|^2 + |a_1|^2 = \int d\theta \ \Psi^*(\theta) \Psi(\theta). \tag{15.12}$$

Since the norm should be a number, and as $\int d\theta \ 1$ is itself a Grassmann variable, the latter should vanish. For the same reason $\int d\theta \ \theta$ (which is itself a commuting object, as is any even product of Grassmann variables) can be seen as a number. Demanding the so-called Grassmann integration to be linear in the integrand, and using $\theta^2 = 0$, all possible ingredients have been discussed. Indeed,

$$\int d\theta \ \theta = 1, \qquad \int d\theta \ 1 = 0 \tag{15.13}$$

is easily seen to give the desired result. Note that $d\theta$ is considered as an independent Grassmann variable (which is important to realise when multiple Grassmann integrations are involved).

We can now study the action of an operator (like a Hamiltonian H) on a state, which in the spinor representation is given by 2×2 matrices.

$$M|\Psi >= |\Psi' > \quad \text{or} \quad \begin{pmatrix} M_{11} & M_{12} \\ M_{21} & M_{22} \end{pmatrix} \begin{pmatrix} a_0 \\ a_1 \end{pmatrix} = \begin{pmatrix} a_0' \\ a_1' \end{pmatrix}. \quad (15.14)$$

Translated to Grassmann variables, this gives

$$\Psi'(\theta) = a_0' + a_1'\theta = M_{11}a_0 + M_{12}a_1 + (M_{21}a_0 + M_{22}a_1)\theta$$
$$\equiv \int d\theta' \, M(\theta, \theta')\Psi(\theta'), \quad (15.15)$$

provided we define

$$M(\theta, \theta') \equiv M_{11}\theta' + M_{12} + M_{21}\theta'\theta - M_{22}\theta$$
$$= M_{11}\theta' + M_{12} - M_{21}\theta\theta' - M_{22}\theta. \quad (15.16)$$

Indeed,

$$\int d\theta' \, M(\theta, \theta')\Psi(\theta') = \int d\theta' \, (M_{11}\theta' + M_{12} - M_{21}\theta\theta' - M_{22}\theta)(a_0 + a_1\theta')$$
$$= \int d\theta' \, \{a_0 M_{12} - a_0 M_{22}\theta + (a_1 M_{12} + a_0 M_{11})\theta'$$
$$- (a_0 M_{21} + a_1 M_{22})\theta\theta'\}$$
$$= \int d\theta' \, \theta'\{(a_1 M_{12} + a_0 M_{11}) + (a_0 M_{21} + a_1 M_{22})\theta\}$$
$$= M_{12}a_1 + M_{11}a_0 + (M_{21}a_0 + M_{22}a_1)\theta. \quad (15.17)$$

The 2×2 identity matrix is hence represented by (note the sign)

$$1_2(\theta, \theta') = \theta' - \theta, \quad (15.18)$$

which can be used to write the infinitesimal evolution operator

$$\exp(-iH\Delta t) = 1_2 - iH\Delta t + \mathcal{O}(\Delta t^2), \quad (15.19)$$

where H is a (possibly time-dependent) 2×2 matrix. In the Grassmann representation, this reads

$$1_2(\theta, \theta') - iH(\theta, \theta')\Delta t = \theta' - \theta - i\Delta t(H_{11}\theta' + H_{12} - H_{21}\theta\theta' - H_{22}\theta)$$
$$= \int d\tilde\theta \, \exp(\tilde\theta[1_2(\theta, \theta') - iH(\theta, \theta')\Delta t]). \quad (15.20)$$

The last identity is exact, and a consequence of the fact that the Taylor series of any function of a Grassmann variable truncates

$$\exp(\tilde\theta x) = 1 + \tilde\theta x, \quad \int d\tilde\theta \, \exp(\tilde\theta x) = x. \quad (15.21)$$

This is valid both for x as a Grassmann variable (in which case the ordering of x with respect to $\tilde{\theta}$ is important, with the opposite ordering the result is $-x$) and for x as a complex number.

Another useful property of Grassmann integration is that (y is a number)

$$\int d\theta \; \exp(\theta x + y) = x \exp(y). \tag{15.22}$$

To prove this, we use that

$$\exp(\theta x + y) = \sum_{n=0}^{\infty} \frac{1}{n!}(\theta x + y)^n = \sum_{n=0}^{\infty} \frac{1}{n!}y^n + \theta x \sum_{n=1}^{\infty} \frac{1}{(n-1)!}y^{n-1}$$
$$= (1 + \theta x) \exp(y). \tag{15.23}$$

In general it is not true that the exponential function retains the property $\exp(x + y) = \exp(x)\exp(y)$, for x and y arbitrary elements of the Grassmann algebra. It behaves as if x and y are matrices, as it should since the Grassmann representation originates from a 2×2 matrix representation. More precisely

$$\exp(\theta x)\exp(\theta' y) = (1 + \theta x)(1 + \theta' y) = 1 + (\theta x + \theta' y) + \tfrac{1}{2}(\theta x + \theta' y)^2$$
$$+ \tfrac{1}{2}[\theta x, \theta' y]$$
$$= \exp\left(\theta x + \theta' y + \tfrac{1}{2}[\theta x, \theta' y]\right). \tag{15.24}$$

This means that the Campbell–Baker–Hausdorff formula [Equation (6.44)] can be extended to this case. It truncates after the single commutator term as neither θx nor θy can appear more than once.

Let us apply this to the evolution operator, which in the Grassmann representation is given by [see Equations (15.19) and (15.20)]

$$< \theta_{i+1}|U(t_{i+1}, t_i)|\theta_i > \; \equiv \; U_i(\theta_{i+1}, \theta_i)$$
$$= \int d\tilde{\theta}_i \; \exp\left(\tilde{\theta}_i[\theta_i - \theta_{i+1}iH(t_i; \theta_{i+1}, \theta_i)\Delta t]\right) + \mathcal{O}(\Delta t^2),$$
$$\tag{15.25}$$

such that

$$\int d\theta_i \; < \theta_{i+1}|U(t_{i+1}, t_i)|\theta_i > < \theta_i|\Psi >$$
$$= \int d\theta_i d\tilde{\theta}_i \; \exp\left(\tilde{\theta}_i[\theta_i - \theta_{i+1} - iH(t_i; \theta_{i+1}, \theta_i)\Delta t]\right) \Psi(\theta_i) + \mathcal{O}(\Delta t^2), \tag{15.26}$$

which can be iterated, first by one step, to

$$
\int d\theta_i < \theta_{i+2}|U(t_{i+2}, t_i)|\theta_i >< \theta_i|\Psi >
$$

$$
\equiv \int d\theta_{i+1} U_{i+1}(\theta_{i+2}, \theta_{i+1}) \int d\theta_i \, U_i(\theta_{i+1}, \theta_i)\Psi(\theta_i)
$$

$$
= \int d\theta_{i+1} d\tilde{\theta}_{i+1} \, \exp\left(\tilde{\theta}_{i+1}[\theta_{i+1} - \theta_{i+2} - i H(t_{i+1}; \theta_{i+2}, \theta_{i+1})\Delta t]\right)
$$

$$
\times \int d\theta_i d\tilde{\theta}_i \, \exp\left(\tilde{\theta}_i[\theta_i - \theta_{i+1} - i H(t_i; \theta_{i+1}, \theta_i)\Delta t]\right) \Psi(\theta_i) + \mathcal{O}(\Delta t^2).
$$

$$(15.27)$$

Note that we have to be careful where we put the differentials, as they are Grassmann variables themselves. If H is diagonal, as will often be the case for the application we have in mind, $\exp\left(\tilde{\theta}[1_2(\theta, \theta') - i H(\theta, \theta')\Delta t]\right)$ will be a commuting object (so-called Grassmann *even*) and it does not matter if we put one of the differentials on one or the other side of the exponential. The combination $d\theta_i d\tilde{\theta}_i$ is likewise Grassmann even, and the pair can be shifted to any place in the expression for the path integral. Hence, provided H is diagonal, any change in the ordering can at most be given an additional minus sign. We now apply Equation (15.24) to the above product of exponentials,

$$
\exp\left(\tilde{\theta}_{i+1}[\theta_{i+1} - \theta_{i+2} - i H(t_{i+1}; \theta_{i+2}, \theta_{i+1})\Delta t]\right)
$$

$$
\times \exp\left(\tilde{\theta}_i[\theta_i - \theta_{i+1} - i H(t_i; \theta_{i+1}, \theta_i)\Delta t]\right)
$$

$$
= \exp\left(\sum_{j=i}^{i+1} \tilde{\theta}_j\{\theta_j - \theta_{j+1} - i H(t_j; \theta_{j+1}, \theta_j)\Delta t\}\right.
$$

$$
\left. - \tfrac{1}{2}\Delta t^2[\tilde{\theta}_{i+1} H(t_{i+1}; \theta_{i+2}, \theta_{i+1}), \tilde{\theta}_i H(t_i; \theta_{i+1}, \theta_i)]\right) \qquad (15.28)
$$

and evaluate the commutator that appears in the exponent. With the explicit expression for H [see Equation (15.20)] we find

$$
\int d\theta_{i+1} \, [\tilde{\theta}_{i+1} H(t_{i+1}; \theta_{i+2}, \theta_{i+1}), \tilde{\theta}_i H(t_i; \theta_{i+1}, \theta_i)]
$$

$$
= 2\tilde{\theta}_{i+1}\tilde{\theta}_i\left(\theta_{i+2} H_{21}(t_{i+1}) H_{12}(t_i) - \theta_i H_{12}(t_{i+1}) H_{21}(t_i)\right). \qquad (15.29)
$$

For H diagonal the commutator term vanishes as was to be expected. In that case, the Campbell–Baker–Hausdorff formula truncates to the trivial term both in the matrix and in the Grassmann representations. This does not mean that there are no discretisation errors in the fermionic path integral when H is diagonal, as can be seen from Equation (15.19). To be precise, assuming for simplicity that $H = -\frac{1}{2}W\sigma_3$, one has

$$\exp(-iH\Delta t) = 1_2 - i\left\{H - \frac{1}{4}W\tan(\frac{1}{4}W\Delta t)1_2\right\}\frac{2\sin(\frac{1}{2}W\Delta t)}{W}, \quad (15.30)$$

which shows that Δt is effectively modified to $2\sin(\frac{1}{2}W\Delta t)/W$ (it is interesting to contrast this with the result we found for the harmonic oscillator in Chapter 6 and Problem 10), whereas H is shifted by a multiple of the identity that vanishes linearly in Δt.

The generalisation of Equations (15.27) and (15.28) to N steps is now obvious and for H diagonal one easily proves that the limit $N \to \infty$ can be taken:

$$\int d\theta \; <\theta'|U(T,0)|\theta> <\theta|\Psi>$$

$$= \lim_{N\to\infty}\int d\theta_0\, U_N(\theta_N = \theta', \theta_0 = \theta)\Psi(\theta_0)$$

$$= \lim_{N\to\infty}\prod_{j=0}^{N-1}\int d\theta_j d\tilde\theta_j\;\exp\left(\sum_{j=0}^{N-1}\tilde\theta_j[\theta_j - \theta_{j+1} - iH(t_j;\theta_{j+1},\theta_j)\Delta t]\right)\Psi(\theta_0),$$

$$(15.31)$$

where, as usual, one has $\Delta t = T/N$. In complete analogy with Equation (6.10), reinstating the dependence on Planck's constant, we can write

$$<\theta'|T\exp(-i\int H(t)dt/\hbar)|\theta>$$

$$= \lim_{N\to\infty}\int d\tilde\theta_0\prod_{j=1}^{N-1}\int d\theta_j d\tilde\theta_j$$

$$\times\exp\left[\frac{i\Delta t}{\hbar}\sum_{j=0}^{N-1}\left(\frac{i\tilde\theta_j(\theta_{j+1} - \theta_j)}{\Delta t} - \tilde\theta_j H(t_j;\theta_{j+1},\theta_j)\right)\right]$$

$$= \lim_{N\to\infty}\int d\tilde\theta_0\prod_{j=1}^{N-1}\int d\theta_j d\tilde\theta_j$$

$$\times\exp\left[\frac{i\Delta t}{\hbar}\sum_{j=0}^{N-1}\left(\frac{i\tilde\theta_j(\theta_{j+1} - \theta_j)}{\Delta t} + \tilde\theta_j W(t_j)\frac{(\theta_{j+1} + \theta_j)}{2}\right)\right]$$

$$\equiv \int D\overline\Psi(t)D\Psi(t)\exp\left[i\int_0^T dt\;\Psi^\dagger(t)(i\partial_t + W(t))\Psi(t)\right]. \quad (15.32)$$

Here we have replaced $\tilde\theta$ with $\Psi^\dagger$ (which in this case agrees with $\overline\Psi$, since in one time and no space dimensions $\gamma_0 \equiv 1$) and θ with Ψ. We have indicated the general case where W can depend on time, but in absence of this time dependence, W is the energy of the one-particle state, created from the vacuum. Since we identified θ also with $b^\dagger$, the creation operator for the one-particle state, we see that $W\Psi^\dagger\Psi = Wbb^\dagger = -H + \frac{1}{2}W$, which is what we expected from a relation between the Lagrangian and Hamiltonian (the term $\frac{1}{2}W$ is of course irrelevant).

As for scalar field theories, we will be mainly interested in vacuum expectation values of the evolution operator. In the presence of a source term, this will allow us to derive all required matrix elements. For the present case we easily find the vacuum wave function to be

$$< \theta|0 >= 1, \qquad < 0|\theta >= \theta, \qquad (15.33)$$

such that

$$< 0|U(T)|0 > = \int d\theta' < 0|\theta' > \int d\theta < \theta'|U(T)|\theta >< \theta|0 >$$

$$= \int d\theta d\theta' \; \theta' < \theta'|U(T)|\theta > . \qquad (15.34)$$

The order of the differentials is important here. However, as in the case of scalar field theories, we do not require the precise form of the vacuum wave function(al) for performing perturbation theory.

Up to now we have described a spin one-half particle pinned-down at a fixed position. It is obvious how this can be generalised to include the quantum mechanical description of a moving spin one-half particle in a one-dimensional potential $V(x)$. If also W depends on the particle position $[W(x)]$, the Hamiltonian becomes

$$H = \left(\frac{\hat{p}^2}{2m} + V(\hat{x}) \right) 1_2 - \tfrac{1}{2}W(\hat{x})\sigma_3. \qquad (15.35)$$

If we write $< x, \theta|\Psi > \equiv \Psi(x, \theta) \equiv a_0(x) + \theta a_1(x)$, the path integral is easily found to be

$$< x', \theta'| \exp(-iHT/\hbar)|x, \theta >$$

$$= \lim_{N\to\infty} \int \frac{dp_0 d\tilde{\theta}_0}{2\pi\hbar} \prod_{j=1}^{N-1} \int \frac{dp_j dx_j}{2\pi\hbar} d\theta_j d\tilde{\theta}_j$$

$$\times \exp\left[\frac{i\Delta t}{\hbar} \sum_{j=0}^{N-1} \left(\frac{p_j(x_{j+1} - x_j)}{\Delta t} - \frac{p_j^2}{2m} - V(x_j) + \frac{i\tilde{\theta}_j(\theta_{j+1} - \theta_j)}{\Delta t} \right. \right.$$

$$\left. \left. + \tilde{\theta}_j W(x_j)\frac{(\theta_{j+1} + \theta_j)}{2} \right) \right]$$

$$\equiv \int \mathcal{D}x(t)\mathcal{D}\overline{\Psi}(t)\mathcal{D}\Psi(t)$$

$$\times \exp\left[\frac{i}{\hbar} \int_0^T dt\left(\tfrac{1}{2}m\dot{x}^2(t) - V(x(t)) + \Psi^\dagger(t)[i\partial_t + W(x(t))]\Psi(t)\right) \right].$$

$$(15.36)$$

A careful derivation of this formula and a step-by-step comparison with the matrix representation in the spin degrees of freedom can be found in Sections 1 to 3 of the paper 'Fermionic Coordinates and Supersymmetry in Quantum Mechanics,' *Nuclear Physics* B196 (1982) 509, by P. Salomonson and J.W. van Holten. For further details, see the lectures by L. Faddeev in 'Methods in Field Theory,' *Les Houches*, 1975, ed. R. Balian and J. Zinn-Justin.

It is now also straightforward to derive the path integral for the Dirac Hamiltonian of the previous chapter. Using as a basis the plane waves constructed there, the Hamiltonian becomes a decoupled sum (in a finite volume) for each $\vec{k}$ of four fermions, described by $b_1^\dagger(\vec{k})$, $b_2^\dagger(\vec{k})$, $d_1^\dagger(\vec{k})$ and $d_2^\dagger(\vec{k})$, each of which can be described by its own θ. Associating the annihilation operators with their respective $\bar{\theta}$ and performing the Fourier transformation back to coordinate space, it is left as an exercise to show that the path integral for fermions is given by

$$\int \mathcal{D}\overline{\Psi}(x)\mathcal{D}\Psi(x) \exp\left[\frac{i}{\hbar} \int d_4x \left(\overline{\Psi}(x)(i\gamma^\mu \partial_\mu - m)\Psi(x)\right.\right.$$
$$\left.\left. - \bar{J}(x)\Psi(x) - \overline{\Psi}(x)J(x)\right)\right]. \tag{15.37}$$

Since the fields $\Psi(x)$ and $\overline{\Psi}(x)$ are Grassmann variables, also the sources $J(x)$ and $\bar{J}(x)$ are Grassmann variables. Their order in the above equation is therefore important when used in further calculations. As promised, it is as simple as for scalar field theories to calculate the dependence of this path integral on the sources. Since Grassmann variables form a complex linear space, we can perform all calculations as in the scalar case, provided we keep track of the order of the Grassmann odd variables. In particular we can make the replacement

$$\Psi(x) \rightarrow \Psi(x) + \int d_4x' \, G_F(x - x')J(x'),$$
$$\overline{\Psi}(x) \rightarrow \overline{\Psi}(x) + \int d_4x' \, \bar{J}(x')G_F(x' - x), \tag{15.38}$$

where $G_F(x)$ is the Green's function defined in Equation (14.9). The integration measure, as for commuting variables, is invariant under a shift by a constant Grassmann variable, such that we obtain (as for the scalar case we normalise the path integral to 1 for vanishing sources)

$$< 0|U_{J\bar{J}}(T)|0 > \; = \; < 0|U_{J=\bar{J}=0}(T)|0 >$$
$$\times \exp\left(-i \int d_4x d_4y \, \bar{J}(x)G_F(x - y)J(y)\right)$$
$$\equiv \; < 0|U_{J=\bar{J}=0}(T)|0 > Z_2(J, \bar{J}). \tag{15.39}$$

Again, this result holds to arbitrary order in the sources, and agrees with what can be derived to second order within Hamiltonian perturbation theory (see Problem 24).

Interactions are taken into account by adding higher-order terms to the Lagrangian, where the order of the fermion fields is important. For example, the Lagrangian for a fermionic and a scalar field is given by

$$\mathcal{L} = \mathcal{L}_2 - V(\overline{\Psi}, \Psi, \sigma) - J(x)\sigma(x) - \bar{J}(x)\Psi(x) - \overline{\Psi}(x)\mathcal{J}(x),$$

$$\mathcal{L}_2 = \overline{\Psi}(x)(i\gamma^\mu \partial_\mu - m)\Psi(x) + \tfrac{1}{2}\partial_\mu\sigma(x)\partial^\mu\sigma(x) - \tfrac{1}{2}M^2\sigma^2(x).$$

$$(15.40)$$

We find as in Equation (8.6)

$$Z(\mathcal{J}, \bar{\mathcal{J}}, J, g_n) = \exp\left(-i \int d_4x \, V\left(\frac{-i\delta}{\delta\mathcal{J}(x)}, \frac{i\delta}{\delta\bar{\mathcal{J}}(x)}, \frac{i\delta}{\delta J(x)}\right)\right) Z_2(\mathcal{J}, \bar{\mathcal{J}}, J),$$

$$Z_2(\mathcal{J}, \bar{\mathcal{J}}, J) = \exp\left[-i \int d_4x d_4y \left(\tfrac{1}{2}J(x)G(x-y)J(y)\right.\right.$$

$$\left.\left. + \bar{\mathcal{J}}(x)G_F(x-y)\mathcal{J}(y)\right)\right].$$

$$(15.41)$$

Note the minus sign for the derivative with respect to $\mathcal{J}(x)$, which is because the source stands behind the field component $\overline{\Psi}(x)$. Derivatives of Grassmann variables are simply defined as one would intuitively expect

$$\frac{d}{d\theta} 1 = 0, \qquad \frac{d}{d\theta} \theta = 1, \qquad \frac{d}{d\theta} \theta' = 0, \qquad (15.42)$$

together with a generalised Leibnitz rule for functions f and g that are either even or odd Grassmann variables, a property denoted by the *sign* or *grading* $s(= \pm 1)$,

$$\frac{d}{d\theta}(fg) = s_f f \frac{d}{d\theta} g + \left(\frac{d}{d\theta} f\right) g. \qquad (15.43)$$

By declaring the derivative to be a linear function on the Grassmann algebra, it can be uniquely extended to this algebra from the above set of rules. Note that these rules imply that the Grassmann integral over a total Grassmann derivative vanishes. Comparing with Equation (15.21) we note that apparently Grassmann integration and differentiation are one and the same thing: Both project on the coefficient in front of the Grassmann variable. The vanishing of the integral over a total derivative and of the derivative of an integral is in that perspective trivial. More importantly, to make sense of Equation (15.41), one easily shows the following identity to hold:

$$\frac{d}{d\theta} \exp(\theta x) = x. \qquad (15.44)$$

An example of an interaction between the fermions and a scalar field σ is given by the so-called Yukawa interaction

$$V(\overline{\Psi}(x), \Psi(x), \sigma(x)) = g\overline{\Psi}(x)\Psi(x)\sigma(x). \qquad (15.45)$$

FIGURE 15.1
Crossing two fermion lines gives a minus sign.

We can also consider the interaction of the fermions with the electromagnetic field, whose quantisation will be undertaken in the next chapter. For this we can take the minimal coupling in Equation (14.12) [see also Equation (14.10)], such that

$$V\big(\overline{\Psi}(x),\, \Psi(x),\, A_\mu(x)\big) = -e\overline{\Psi}(x)\gamma^\mu \Psi(x) A_\mu(x), \qquad (15.46)$$

which will play a dominating role in describing quantum electrodynamics (QED).

As before, $\log Z(\bar{\mathcal{J}}, \mathcal{J}, J)$ is the sum over connected diagrams. Diagrams that involve fermions necessarily have as many lines ending in a source $\mathcal{J}$ as in a source $\bar{\mathcal{J}}$. This is because the Lagrangian is Grassmann even, a requirement that can be related to the Lorentz invariance. It does not in general require the Lagrangian to be bilinear in $\overline{\Psi}$ and Ψ. In Chapter 19 and Problem 30, we will discuss the four-Fermi interaction, $\overline{\Psi}(x)\gamma^\mu\Psi(x)\overline{\Psi}(x)\gamma_\mu\Psi(x)$, which is clearly Lorentz invariant and Grassmann even. However, for many of the theories we discuss, the Lagrangian is bilinear in the fermionic fields, because higher-order terms will generally not be renormalisable (except in one space and one time dimension). If no higher-order fermionic interactions occur, a fermionic line either forms a loop or it goes from a source $\mathcal{J}$ to a source $\bar{\mathcal{J}}$. As changing the order of fermionic fields and sources gives a sign change, this has consequences for the Feynman diagrams too. However, it is cumbersome to determine the overall sign of a diagram. Fortunately, all we need is the relative sign of the various diagrams that contribute to the Green's function with a fixed number of sources, since the overall sign drops out in our computations of cross sections and decay rates. If one diagram can be obtained from the other by crossing two fermion lines, this gives a relative minus sign, as for Figure 15.1.

It also implies that each loop formed by a fermion line carries a minus sign. Intuitively this follows, as is indicated in Figure 15.2 by the dashed box, from the identity displayed in Figure 15.1.

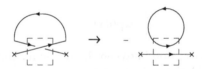

FIGURE 15.2
Fermion loops carry a minus sign.

More accurately a fermion loop that connects vertices x_k for $k = 1$ to n is associated to

$$\prod_{k=1}^{n} \overline{\Psi}(x_k)\Psi(x_k) \rightarrow \prod_{k=1}^{n} \left(\frac{-i\delta}{\delta \mathcal{J}(x_k)} \frac{i\delta}{\delta \overline{\mathcal{J}}(x_k)} \right) \rightarrow$$

$$- \sum_{\{k\}} \text{Tr}\left[iG_F(x_{k(1)} - x_{k(2)}) iG_F(x_{k(2)} - x_{k(3)}) \cdots iG_F(x_{k(n)} - x_{k(1)}) \right],$$

(15.47)

where $\{k\}$ stands for the various orders in which the vertices are connected. For each vertex we have only indicated the fermionic part $\overline{\Psi}(x_k)\Psi(x_k)$. For the examples of Equations (15.45) and (15.46), the scalar or vector field contributions are not indicated, as they are not relevant for the fermion loop. The trace is with respect to the spinor indices, which are not displayed explicitly to keep the notation simple. We used that

$$\Psi(x_{k(j)})\overline{\Psi}(x_{k(j+1)}) \rightarrow \frac{i\delta}{\delta \overline{\mathcal{J}}(x_{k(j)})} \frac{-i\delta}{\delta \mathcal{J}(x_{k(j+1)})} \rightarrow iG_F(x_{k(j)} - x_{k(j+1)}), \quad (15.48)$$

where an extra minus sign arises since in $\log Z_2(\mathcal{J}, \overline{\mathcal{J}})$ [see Equation (15.39)] $\overline{\mathcal{J}}$ comes first, and has to be anticommuted with $\delta/\delta\mathcal{J}$ before this derivative can be taken. The overall minus sign comes from the term that closes the loop

$$\overline{\Psi}(x_{k(1)})A\Psi(x_{k(n)}) = -A\Psi(x_{k(n)})\overline{\Psi}(x_{k(1)}), \quad (15.49)$$

where A is Grassmann even. We contrast this, as an example, with a scalar loop for the field φ that arises in the theory discussed in Chapter 11, which is described by the Lagrangian $\mathcal{L} = \frac{1}{2}(\partial_\mu\varphi)^2 - \frac{1}{2}m^2\varphi^2 + \frac{1}{2}(\partial_\mu\sigma)^2 - \frac{1}{2}M^2\sigma^2 - \frac{1}{2}\varphi^2\sigma - J\sigma - j\varphi$. One finds $[k(n+1) \equiv k(1)]$

$$\prod_{k=1}^{n} \varphi(x_k)\varphi(x_k) \rightarrow \prod_{k=1}^{n} \left(\frac{i\delta}{\delta J(x_k)} \frac{i\delta}{\delta J(x_k)} \right) \rightarrow \sum_{\{k\}} \prod_{j=1}^{n} iG(x_{k(j)} - x_{k(j+1)}), \quad (15.50)$$

which completes the demonstration of the extra minus sign for fermion loops. Note that the factors of i, associated to the derivatives with respect to the sources, are absorbed in the vertices for the Feynman rules in Table 8.1 on page 53, which is why the propagator in that table equals $-i$ times the Green's function. This is also the Feynman rule for the fermion propagator. But the extra minus sign in the derivative with respect to the fermionic source $\mathcal{J}$ [see Equation (15.41)] is *not* absorbed in the vertex in order to guarantee that vertex factors are assigned as in the scalar theory. That minus sign is, however, absorbed in Equation (15.48) due to the anticommuting nature of the fermionic variables, which was in the first place the reason for the extra minus sign in Equation (15.41) to appear. Only the overall minus sign required in closing a fermion loop remains as an extra factor.

100 A Course in Field Theory

Before we convert these Feynman rules to the ones involved in computing the scattering matrix, cross sections, and decay rates [see pg. 63 and Equation (9.25)], we need to determine the wave-function factors to be inserted for the external lines that correspond to the in- and outgoing fermion lines. For this we express the creation and annihilation operators in terms of the fermionic fields (at $t = 0$), such that their insertion in the operator formulation can be converted in the path integral, as in the scalar case, to derivatives with respect to the sources. Using Equation (13.17) and the orthogonality relations of Equation (13.14), one finds [compare this to Equation (7.26)]

$$b_\alpha(\vec{k}) = \frac{u^{(\alpha)}(\vec{k})^\dagger}{\sqrt{2k_0(\vec{k})}} \int \frac{d_3\vec{x}}{\sqrt{(2\pi)^3}} e^{-i\vec{k}\cdot\vec{x}} \hat{\Psi}(\vec{x}),$$

$$d_\alpha^\dagger(\vec{k}) = \frac{v^{(\alpha)}(\vec{k})^\dagger}{\sqrt{2k_0(\vec{k})}} \int \frac{d_3\vec{x}}{\sqrt{(2\pi)^3}} e^{i\vec{k}\cdot\vec{x}} \hat{\Psi}(\vec{x}),$$

(15.51)

and through conjugation we get

$$b_\alpha^\dagger(\vec{k}) = \int \frac{d_3\vec{x}}{\sqrt{(2\pi)^3}} e^{i\vec{k}\cdot\vec{x}} \hat{\Psi}^\dagger(\vec{x}) \frac{u^{(\alpha)}(\vec{k})}{\sqrt{2k_0(\vec{k})}},$$

$$d_\alpha(\vec{k}) = \int \frac{d_3\vec{x}}{\sqrt{(2\pi)^3}} e^{-i\vec{k}\cdot\vec{x}} \hat{\Psi}^\dagger(\vec{x}) \frac{v^{(\alpha)}(\vec{k})}{\sqrt{2k_0(\vec{k})}}.$$

(15.52)

In the Hamiltonian formulation the scattering matrix is given by

$$_{out}< (\vec{p}_1, \alpha_1), (\vec{p}_2, \alpha_2), \ldots, (\vec{p}_\ell, \alpha_\ell)|(\vec{k}_1, \beta_1), (\vec{k}_2, \beta_2), \cdots, (\vec{k}_n, \beta_n) >_{in}$$
$$= < 0|c_{\alpha_1}(\vec{p}_1)c_{\alpha_2}(\vec{p}_2)\cdots c_{\alpha_\ell}(\vec{p}_\ell)U(T_{out}, T_{in})c_{\beta_1}^\dagger(\vec{k}_1)c_{\beta_2}^\dagger(\vec{k}_2)\cdots c_{\beta_n}^\dagger(\vec{k}_n)|0 >,$$

(15.53)

where in a shorthand notation we separate particles from antiparticles by the helicity index

$$c_1(\vec{k}) \equiv b_1(\vec{k}), \quad c_2(\vec{k}) \equiv b_2(\vec{k}), \quad c_3(\vec{k}) \equiv d_1(\vec{k}), \quad c_4(\vec{k}) \equiv d_2(\vec{k}). \quad (15.54)$$

Like in Equation (9.3), the insertion of a field operator at the appropriate time is in the path integral represented by a derivative with respect to the source

$$\hat{b}_\alpha^\dagger(\vec{k}) = \frac{\left[\gamma^0 u^{(\alpha)}(\vec{k})\right]_a}{\sqrt{2k_0(\vec{k})(2\pi)^3}} \int d_3\vec{x}\, e^{i\vec{k}\cdot\vec{x}} \frac{-i\delta}{\delta J_a(\vec{x}, t = T_{in})}$$

$$= \frac{\left[\gamma^0 u^{(\alpha)}(\vec{k})\right]_a}{\sqrt{2k_0(\vec{k})}} \frac{-i\delta}{\delta \tilde{J}_a(\vec{k}, t = T_{in})},$$

$$\hat{a}_\alpha^\dagger(\vec{k}) = \frac{v_a^{(\alpha)}(\vec{k})^*}{\sqrt{2k_0(\vec{k})(2\pi)^3}} \int d_3\vec{x}\, e^{i\vec{k}\cdot\vec{x}} \frac{i\delta}{\delta \bar{\mathcal{J}}_a(\vec{x}, t = T_{\text{in}})} = \frac{v_a^{(\alpha)}(\vec{k})^*}{\sqrt{2k_0(\vec{k})}} \frac{i\delta}{\delta \bar{\tilde{\mathcal{J}}}_a(\vec{k}, t = T_{\text{in}})},$$

$$\hat{b}_\alpha(\vec{k}) = \frac{u_a^{(\alpha)}(\vec{k})^*}{\sqrt{2k_0(\vec{k})(2\pi)^3}} \int d_3\vec{x}\, e^{-i\vec{k}\cdot\vec{x}} \frac{i\delta}{\delta \bar{\mathcal{J}}_a(\vec{x}, t = T_{\text{out}})}$$

$$= \frac{u_a^{(\alpha)}(\vec{k})^*}{\sqrt{2k_0(\vec{k})}} \frac{i\delta}{\delta \bar{\tilde{\mathcal{J}}}_a(-\vec{k}, t = T_{\text{out}})},$$

$$\hat{a}_\alpha(\vec{k}) = \frac{\left[\gamma^0 v^{(\alpha)}(\vec{k})\right]_a}{\sqrt{2k_0(\vec{k})(2\pi)^3}} \int d_3\vec{x}\, e^{-i\vec{k}\cdot\vec{x}} \frac{-i\delta}{\delta \mathcal{J}_a(\vec{x}, t = T_{\text{out}})}$$

$$= \frac{\left[\gamma^0 v^{(\alpha)}(\vec{k})\right]_a}{\sqrt{2k_0(\vec{k})}} \frac{-i\delta}{\delta \tilde{\mathcal{J}}_a(-\vec{k}, t = T_{\text{out}})},$$

$$\tag{15.55}$$

such that

$$_{\text{out}}< (\vec{p}_1, \alpha_1), (\vec{p}_2, \alpha_2), \dots, (\vec{p}_\ell, \alpha_\ell)|(\vec{k}_1, \beta_1), (\vec{k}_2, \beta_2), \dots, (\vec{k}_n, \beta_n) >_{\text{in}}$$

$$= \prod_{i=1}^{\ell} \hat{c}_{\alpha_i}(\vec{p}_i) \prod_{j=1}^{n} \hat{c}_{\beta_j}^\dagger(\vec{k}_j)\, \exp(G_{\mathcal{J}, \bar{\mathcal{J}}})|_{\mathcal{J}=\bar{\mathcal{J}}=0}. \tag{15.56}$$

The $\hat{c}_\alpha(\vec{k})$ are, of course, defined in terms of $\hat{b}_\alpha(\vec{k})$ and $\hat{d}_\alpha(\vec{k})$ as in Equation (15.54). We continue as in Chapter 9 by first fixing the wave-function and mass renormalisations in terms of the connected two-point function. (We leave it as an exercise to show that in the absence of interactions $_{\text{out}}< (\vec{p}, \alpha)|(\vec{k}, \beta) >_{\text{in}} = e^{-ip_0(\vec{p})T} \delta_{\vec{k}, \vec{p}} \delta_{\alpha, \beta}.$)

$$G_c^{(2)}(\mathcal{J}, \bar{\mathcal{J}}) = \times\!\!\longrightarrow\!\!\times \; + \; \times\!\!\rightarrow\!\!\bigcirc\!\!\rightarrow\!\!\times \; + \; \times\!\!\rightarrow\!\!\bigcirc\!\!\rightarrow\!\!\bigcirc\!\!\rightarrow\!\!\times \; + \cdots$$

$$\equiv \; \times\!\!\longrightarrow\!\!\bigotimes\!\!\longrightarrow\!\!\times, \tag{15.57}$$

where the self-energy is now a 4×4 matrix given by $(-i\times)$ the amputated $1PI$ two-point function

$$i\Sigma_{ab}(p) \equiv \;\;\; \underset{b}{\longrightarrow}\!\!\bigcirc\!\!\underset{a}{\longrightarrow}\;. \tag{15.58}$$

(The $1PI$ diagram *equals* $\Sigma_{ab}(p)$, when evaluated with the Feynman rules for the reduced matrix elements of Table 15.1 (pg. 106) by *dropping* the wave-function factors.) The convention for these sort of diagrams is that momentum flows in the direction of the arrow, which points to the *first* spinor index (here a). With these definitions the two-point function becomes

$$G_c^{(2)}(\mathcal{J}, \bar{\mathcal{J}}) = -i \int d_4 p\, \bar{\mathcal{J}}_a(-p) \left[\frac{1}{\not{p} - m - \Sigma(p) + i\varepsilon} \right]_{ab} \bar{\mathcal{J}}_b(p), \qquad (15.59)$$

with between square brackets the inverse of the 4×4 matrix $[p_\mu \gamma^\mu - m - \Sigma(p) + i\varepsilon]_{ab}$, which is the full propagator in the momentum representation for the conventions on page. 63. As long as we don't break the Lorentz invariance, the full propagator near the poles is of the form of the free propagator with a wave-function renormalisation factor Z_F and a renormalised mass $\tilde{m}$, such that on the mass-shell one has [compare this to Equation (9.12)]

$$G_c^{(2)}(\mathcal{J}, \bar{\mathcal{J}}) = -i Z_F \int d_4 p\, \bar{\mathcal{J}}(-p) \left[\frac{1}{\not{p} - \tilde{m} + i\varepsilon} \right] \mathcal{J}(p). \qquad (15.60)$$

Performing the wave-function renormalisation, we have to modify Equations (15.55) and (15.56) accordingly [compare these to Equations (9.13) and (9.14)].

$$b_+^\alpha(\vec{k}) = \frac{\left[\gamma^0 u^{(\alpha)}(\vec{k}) \right]_a}{\sqrt{2 Z_F \sqrt{\vec{k}^2 + \tilde{m}^2}}} \frac{-i\delta}{\delta \bar{\mathcal{J}}_a(\vec{k}, t = T_{\text{in}})},$$

$$b_-^\alpha(\vec{k}) = \frac{u_a^{(\alpha)}(\vec{k})^*}{\sqrt{2 Z_F \sqrt{\vec{k}^2 + \tilde{m}^2}}} \frac{i\delta}{\delta \mathcal{J}_a(-\vec{k}, t = T_{\text{out}})},$$

$$d_+^\alpha(\vec{k}) = \frac{v_a^{(\alpha)}(\vec{k})^*}{\sqrt{2 Z_F \sqrt{\vec{k}^2 + \tilde{m}^2}}} \frac{i\delta}{\delta \bar{\mathcal{J}}_a(\vec{k}, t = T_{\text{in}})}, \qquad (15.61)$$

$$d_-^\alpha(\vec{k}) = \frac{\left[\gamma^0 v^{(\alpha)}(\vec{k}) \right]_a}{\sqrt{2 Z_F \sqrt{\vec{k}^2 + \tilde{m}^2}}} \frac{-i\delta}{\delta \mathcal{J}_a(-\vec{k}, t = T_{\text{out}})},$$

and, with the $\hat{c}_\pm^\alpha(\vec{k})$ defined as in Equation (15.54),

$$_{\text{out}}< (\vec{p}_1, \alpha_1), (\vec{p}_2, \alpha_2), \ldots, (\vec{p}_\ell, \alpha_\ell) | (\vec{k}_1, \beta_1), (\vec{k}_2, \beta_2), \ldots, (\vec{k}_n, \beta_n) >_{\text{in}}$$

$$= \prod_{i=1}^{\ell} \hat{c}_-^{\alpha_i}(\vec{p}_i) \prod_{j=1}^{n} \hat{c}_+^{\beta_j}(\vec{k}_j) \exp(G_{\mathcal{J}, \bar{\mathcal{J}}})|_{\mathcal{J} = \bar{\mathcal{J}} = 0}. \qquad (15.62)$$

To compute the wave-function factors for the external lines, we express the n-point function in terms of amputated n-point functions, as in Equation (9.15), with the difference that there has to be an even number of external lines, since the number of $\mathcal{J}$ and $\bar{\mathcal{J}}$ sources has to be equal (we ignore for the moment any other bosonic fields that might be present, including those in the external lines that will be obvious). The amputated n-point function will now carry the spinor index of each of the external lines and one has

$$G_c^{(2n)}(\mathcal{J}, \tilde{\mathcal{J}}) \equiv \int \prod_{j=1}^{n} d_4 p_j d_4 k_j \, G_c^{(\text{amp})}(p_1, p_2, \ldots, p_n; k_1, k_2, \ldots, k_n)_{a_1,a_2,\ldots,a_n}^{b_1,b_2,\ldots,b_n}$$

$$\times \prod_{j=1}^{n} \left\{ \tilde{\mathcal{J}}_{c_j}(p_j) \left[\frac{-i}{-\not{p}_j - m - \Sigma(-p_j) + i\varepsilon} \right]_{c_j b_j} \right.$$

$$\times \left. \left[\frac{-i}{\not{k}_j - m - \Sigma(k_j) + i\varepsilon} \right]_{a_j d_j} \tilde{\mathcal{J}}_{d_j}(k_j) \right\}. \tag{15.63}$$

Note that we have assumed one particular ordering for the sources. Relative signs of the diagrams are determined by the rules that were described above (it is not difficult to convince oneself that with respect to the fermion lines, any diagram can be generated from a given one by permuting fermion lines). As in Equations (9.20) and (9.21), we can compute the action of $\hat{c}_{\pm}^{\alpha}(\vec{k})$, from which the wave-function factors will be obtained. Like in Equation (9.19), computing the action of $\hat{b}_{-}^{\alpha}(\vec{p})$ and $\hat{a}_{+}^{\alpha}(\vec{p})$ we can restrict our attention to

$$A(\cdots)_{\cdots}^{\cdots} \equiv \int d_4 p \, \tilde{\mathcal{J}}_a(p) \left[\frac{-i}{-\not{p} - m - \Sigma(-p) + i\varepsilon} \right]_{ab} G_c^{(\text{amp})}(p, \ldots)_{\cdots}^{b\cdots}$$

$$= \int \frac{d_4 p}{\sqrt{2\pi}} \int dt \, \tilde{\mathcal{J}}_a(\vec{p}, t) \left[\frac{-i e^{i p_0 t}}{-\not{p} - m - \Sigma(-p) + i\varepsilon} \right]_{ab}$$

$$\times G_c^{(\text{amp})}(p, \ldots)_{\cdots}^{b\cdots}. \tag{15.64}$$

If we define as usual $p_0(\vec{p}) = \sqrt{\vec{p}^2 + \tilde{m}^2}$ and for convenience of notation change p to $-p$ in Equation (15.64), we find

$$\hat{b}_{-}^{\alpha}(\vec{p}) A(\cdots)_{\cdots}^{\cdots} = \frac{u_a^{(\alpha)}(\vec{p})^*}{\sqrt{4\pi p_0(\vec{p}) Z_F}} \int dp_0 \left[\frac{e^{-i p_0 T_{\text{out}}}}{\not{p} - m - \Sigma(p) + i\varepsilon} \right]_{ab}$$

$$\times G_c^{(\text{amp})}(-p, \ldots)_{\cdots}^{b\cdots}$$

$$= -i\sqrt{(2\pi)^4 Z_F} \, u_a^{(\alpha)}(\vec{p})^* \frac{(\not{p} + \tilde{m})_{ab}}{2 p_0(\vec{p})} \frac{G_c^{(\text{amp})}(-p, \ldots)_{\cdots}^{b\cdots}}{\sqrt{2 p_0(\vec{p})(2\pi)^3}}$$

$$\times e^{-i p_0(\vec{p}) T_{\text{out}}}, \tag{15.65}$$

which is obtained by deforming the p_0 integration contour to the lower half-plane (since $T_{\text{out}} \to \infty$), and computing the contribution from the pole at $p_0 = p_0(\vec{p})$, taking into account Equation (15.60). Its residue is proportional to the matrix $Z_F(\not{p} + \tilde{m})/2 p_0(\vec{p})$, which is most easily found using $1/(\not{p} - \tilde{m} + i\varepsilon) =$

$(\not{p}+\tilde{m})/(p^2-\tilde{m}^2+i\varepsilon)$. We can now use the fact that the spinor $u^{(\alpha)}(\vec{p})$ satisfies the equations of motion [see Equation (13.2)], such that

$$u_a^{(\alpha)}(\vec{p})^* \frac{(\not{p}+\tilde{m})_{ab}}{2p_0(\vec{p})} = \left(\frac{(\not{p}+\tilde{m})^\dagger}{2p_0(\vec{p})} u^{(\alpha)}(\vec{p})\right)_b^* = \left(\frac{(\gamma^0\not{p}\gamma^0+\tilde{m})}{2p_0(\vec{p})} u^{(\alpha)}(\vec{p})\right)_b^*$$

$$= \overline{u^{(\alpha)}(\vec{p})}_b. \tag{15.66}$$

Consequently, Equation(15.65) becomes

$$\hat{b}_-^\alpha(\vec{p}) A(\cdots)_{\cdots}^{\cdots} = -i\overline{u^{(\alpha)}(\vec{p})}_b \sqrt{(2\pi)^4 Z_F} \frac{G_c^{(\mathrm{amp})}(-p,\dots)_{\cdots}^{b\cdots}}{\sqrt{2p_0(\vec{p})(2\pi)^3}} e^{-ip_0(\vec{p})T_{\mathrm{out}}}. \tag{15.67}$$

The wave-function factor for an *outgoing* electron is therefore given by $\overline{u^{(\alpha)}(\vec{p})}_b \sqrt{Z_F}$. The convention is that the momentum flows out of the diagram, along the arrow (see Table 15.1); this is why the amputated n-point function has $-p$ as its argument, like for the scalar case, where we defined for the amputated n-point function all momenta to flow into the diagram. This means that in the reduced matrix element $\mathcal{M}$, the outgoing electron momenta occur precisely as indicated in Equation (9.18).

By similar arguments we obtain from Equation (15.64)

$$\hat{d}_+^\alpha(\vec{p}) A(\cdots)_{\cdots}^{\cdots} = \frac{v_a^{(\alpha)}(\vec{p})^*}{\sqrt{4\pi p_0(\vec{p})Z_F}} \int dp_0 \left[\frac{e^{ip_0 T_{\mathrm{in}}}}{-\not{p}-m-\Sigma(-p)+i\varepsilon}\right]_{ab}$$

$$\times G_c^{(\mathrm{amp})}(p,\dots)_{\cdots}^{b\cdots}$$

$$= -i\sqrt{(2\pi)^4 Z_F}\, v_a^{(\alpha)}(\vec{p})^* \frac{(-\not{p}+\tilde{m})_{ab}}{2p_0(\vec{p})} \frac{G_c^{(\mathrm{amp})}(p,\dots)_{\cdots}^{b\cdots}}{\sqrt{2p_0(\vec{p})(2\pi)^3}} e^{ip_0(\vec{p})T_{\mathrm{in}}}$$

$$= i\overline{v^{(\alpha)}(\vec{p})}_b \sqrt{(2\pi)^4 Z_F} \frac{G_c^{(\mathrm{amp})}(p,\dots)_{\cdots}^{b\cdots}}{\sqrt{2p_0(\vec{p})(2\pi)^3}} e^{ip_0(\vec{p})T_{\mathrm{in}}}, \tag{15.68}$$

such that the wave-function factor for an *incoming* antiparticle (positron) is $-\overline{v^{(\alpha)}(\vec{p})}_b \sqrt{Z_F}$. In this case the momentum flows *against* the arrow of the fermion line, but does flow into the diagram as is required in the convention of the reduced matrix element.

To compute the action of $\hat{b}_+^\alpha(\vec{p})$ and $\hat{d}_-^\alpha(\vec{p})$, we restrict our attention to

$$B(\cdots)_{\cdots}^{\cdots} \equiv \int d_4 p\, G_c^{(\mathrm{amp})}(p,\dots)_{b\cdots}^{\cdots} \left[\frac{-i}{\not{p}-m-\Sigma(p)+i\varepsilon}\right]_{ba} \tilde{J}_a(p)$$

$$= \int \frac{d_4 p}{\sqrt{2\pi}} \int dt\, G_c^{(\mathrm{amp})}(p,\dots)_{b\cdots}^{\cdots} \left[\frac{-ie^{ip_0 t}}{\not{p}-m-\Sigma(p)+i\varepsilon}\right]_{ba} \tilde{J}_a(\vec{p},t). \tag{15.69}$$

One finds

$$
\hat{b}_+^\alpha(\vec{p}) B(\cdots)_{\cdots}^{\cdots} = \left\{ \int dp_0 G_c^{(amp)}(p,\dots)_{b\cdots}^{\cdots} \left[\frac{e^{ip_0 T_{in}}}{\not{p} - m - \Sigma(p) + i\varepsilon} \right]_{ba} \right\}
$$

$$
\times \frac{[\gamma^0 u^{(\alpha)}(\vec{p})]_a}{\sqrt{4\pi p_0(\vec{p}) Z_F}}
$$

$$
= -i\sqrt{(2\pi)^4 Z_F} \, \frac{G_c^{(amp)}(p,\dots)_{b\cdots}^{\cdots}}{\sqrt{2p_0(\vec{p})(2\pi)^3}} \, \frac{(\not{p} + \tilde{m})_{ba}}{2p_0(\vec{p})}
$$

$$
\times [\gamma^0 u^{(\alpha)}(\vec{p})]_a \, e^{ip_0(\vec{p}) T_{in}}
$$

$$
= -i\sqrt{(2\pi)^4 Z_F} \, \frac{G_c^{(amp)}(p,\dots)_{b\cdots}^{\cdots}}{\sqrt{2p_0(\vec{p})(2\pi)^3}} u_b^{(\alpha)}(\vec{p}) e^{ip_0(\vec{p}) T_{in}}, \quad (15.70)
$$

and (again for convenience changing p to $-p$)

$$
\hat{d}_-^\alpha(\vec{p}) B(\cdots)_{\cdots}^{\cdots} = \left\{ \int dp_0 G_c^{(amp)}(-p,\dots)_{b\cdots}^{\cdots} \left[\frac{e^{-ip_0 T_{out}}}{-\not{p} - m - \Sigma(-p) + i\varepsilon} \right]_{ba} \right\}
$$

$$
\times \frac{[\gamma^0 v^{(\alpha)}(\vec{p})]_a}{\sqrt{4\pi p_0(\vec{p}) Z_F}}
$$

$$
= -i\sqrt{(2\pi)^4 Z_F} \, \frac{G_c^{(amp)}(-p,\dots)_{b\cdots}^{\cdots}}{\sqrt{2p_0(\vec{p})(2\pi)^3}} \, \frac{(-\not{p} + \tilde{m})_{ba}}{2p_0(\vec{p})}
$$

$$
\times [\gamma^0 v^{(\alpha)}(\vec{p})]_a \, e^{-ip_0(\vec{p}) T_{in}}
$$

$$
= i\sqrt{(2\pi)^4 Z_F} \, \frac{G_c^{(amp)}(-p,\dots)_{b\cdots}^{\cdots}}{\sqrt{2p_0(\vec{p})(2\pi)^3}} \, v_b^{(\alpha)}(\vec{p}) e^{-ip_0(\vec{p}) T_{out}}. \quad (15.71)
$$

In both cases there is an extra minus sign from pulling $\delta/\delta J$ through $\bar{J}$ in Equation (15.63). The wave-function factor for an *incoming* electron is hence $u_b^{(\alpha)}(\vec{p})\sqrt{Z_F}$ with the momentum flowing along the fermionic arrow, whereas the wave-function factor of an *outcoming* antiparticle (or positron) is given by $-v_b^{(\alpha)}(\vec{p})\sqrt{Z_F}$, where the momentum flows opposite to the fermionic arrow. The minus signs in front of some of the wave-function factors are irrelevant (they can be absorbed in the overall sign ambiguity).

In Table 15.1 we summarise the Feynman rules that correspond to the fermionic pieces in computing the reduced matrix elements. We have chosen the convention that the incoming momenta flow in, and the outgoing momenta flow out of the diagram. This guarantees that Equations (9.18), (10.12), and (11.11) (resp. the scattering matrix, cross section, and decay rate) remain valid in the presence of fermions. Consequently, all fermion momenta in the

TABLE 15.1

Feynman rules for fermions.

	$\equiv$ $g\delta_{ab}$ and $k_1 = k_2 + k_3$	Yukawa vertex
	$\equiv$ $-e\gamma^\mu_{ab}$ and $k_1 = k_2 + k_3$	photon vertex
$\overrightarrow{b\ k\ a}$	$\equiv$ $\left[\frac{1}{\not{k}-m+i\varepsilon}\right]_{ab}$	fermion propagator
	$\equiv$ $\sqrt{Z_F}\, u_a^{(\alpha)}(\vec{k})$, $k_0 = \sqrt{\vec{k}^2+\bar{m}^2}$	incoming electron
	$\equiv$ $\sqrt{Z_F}\, \overline{v^{(\alpha)}}(\vec{k})_a$, $k_0 = \sqrt{\vec{k}^2+\bar{m}^2}$	incoming positron
	$\equiv$ $\sqrt{Z_F}\, \overline{u^{(\alpha)}}(\vec{k})_a$, $k_0 = \sqrt{\vec{k}^2+\bar{m}^2}$	outgoing electron
	$\equiv$ $\sqrt{Z_F}\, v_a^{(\alpha)}(\vec{k})$, $k_0 = \sqrt{\vec{k}^2+\bar{m}^2}$	outgoing positron
	$-1 \times i \int \frac{d_4 k}{(2\pi)^4}$	for each *fermion* loop
	-1	interchange of *fermion* lines

table flow from left to right. For conventions where momenta always flow in the direction of the fermion arrow, the four momenta for wave-function factors associated to in- and outgoing antiparticles (positrons) should be reversed. Signs from fermion loops and exchanges of external lines will *not* be implicit in diagrams, as only *relative* signs are known.

16

Feynman Rules for Vector Fields

DOI: 10.1201/b15364-16

As before, the quantisation for vector fields starts by expanding the field in plane waves and identifying the Fourier coefficients with creation and annihilation operators:

$$A_\mu(x) = \int \frac{d_3\vec{k}}{\sqrt{2k_0(\vec{k})(2\pi)^3}} \sum_\lambda \left(a_\lambda(\vec{k})\varepsilon_\mu^{(\lambda)}(\vec{k})e^{-ikx} + a_\lambda^\dagger(\vec{k})\varepsilon_\mu^{(\lambda)}(\vec{k})^* e^{ikx}\right), \quad (16.1)$$

where $\varepsilon_\mu^{(\lambda)}(\vec{k})e^{-ikx}$ are independent plane wave solutions of the equations of motion. The index λ enumerates the various solutions for fixed momentum. They will be identified with the spin components or helicity eigenstates of the vector.

We will first discuss the simpler case of a massive vector field, expected to describe a massive spin-one particle. In Problem 12 we already saw that its Lagrangian is given by

$$\mathcal{L}_A = -\tfrac{1}{4}F_{\mu\nu}F^{\mu\nu} + \tfrac{1}{2}m^2 A_\mu A^\mu - A_\mu(x)J^\mu(x), \quad (16.2)$$

and that the free equations of motion (i.e., $J^\mu = 0$) are equivalent to

$$\partial_\mu A^\mu(x) = 0 \quad \text{and} \quad (\partial_\mu\partial^\mu + m^2)A_\nu(x) = 0. \quad (16.3)$$

As usual, this implies the on-shell condition $k_0^2 = \vec{k}^2 + m^2$, but also

$$k^\mu \varepsilon_\mu^{(\lambda)}(\vec{k}) = 0. \quad (16.4)$$

It removes one of the four degrees of freedom of a four-vector. Three independent components remain, exactly what would be required for a particle with spin one. We may, for example, choose

$$\varepsilon_\mu^{(\lambda)}(\vec{0}) = \delta_\mu^\lambda \quad (\lambda = 1, 2, 3) \quad (16.5)$$

in the rest frame of the particle, which is extended to an arbitrary frame by applying a Lorentz boost. They satisfy the Lorentz invariant normalisation

$$g^{\mu\nu}\varepsilon_\mu^{(\lambda)}(\vec{k})^*\varepsilon_\nu^{(\lambda')}(\vec{k}) = -\delta^{\lambda\lambda'}. \quad (16.6)$$

The minus sign is just a consequence of the fact that in our conventions $g_{ij} = -\delta_{ij}$. The spin wave functions also satisfy a completeness relation which is given by

$$\Lambda_{\mu\nu}(\vec{k}) \equiv \sum_\lambda \varepsilon_\mu^{(\lambda)}(\vec{k})\varepsilon_\nu^{(\lambda)}(\vec{k})^* = -\left(g_{\mu\nu} - \frac{k_\mu k_\nu}{m^2}\right), \qquad (16.7)$$

most easily proven in the rest frame. Since the spin wave functions are by construction Lorentz vectors, the above expression forms a Lorentz tensor and its *on-shell* extension to an arbitrary frame is therefore unique. It is easily seen to project arbitrary vectors w_μ on to vectors that satisfy $k^\mu w_\mu = 0$. The propagator for the massive spin-one field was already computed in Problem 12:

$$
\begin{array}{c}
k \\[-2pt]
\mu \;\rule{1.2cm}{0pt}\; \nu
\end{array}
= \frac{-\left(g_{\mu\nu} - \frac{k_\mu k_\nu}{m^2}\right)}{k^2 - m^2 + i\varepsilon} = \frac{\Lambda_{\mu\nu}(\vec{k})}{k^2 - m^2 + i\varepsilon}. \qquad (16.8)
$$

Especially for the massless spin-one field (i.e., the photon field) to be discussed below, it is advantageous to decompose the spin with respect to the direction of the particle's motion, which are called helicity eigenstates. We have helicities 0 and ± 1, described by the spin wave functions

$$\varepsilon_\mu^{(0)}(k) = \left(\frac{|\vec{k}|}{m}, \frac{-k_0}{m|\vec{k}|}\vec{k}\right), \quad \varepsilon_\mu^{(\pm)}(k) = \tfrac{1}{2}\sqrt{2}\left(\bar{\varepsilon}_\mu^{(1)}(k) \pm i\bar{\varepsilon}_\mu^{(2)}(k)\right), \qquad (16.9)$$

where k_μ, $\varepsilon_\mu^{(0)}(k)$, $\bar{\varepsilon}_\mu^{(1)}(k)$ and $\bar{\varepsilon}_\mu^{(2)}(k)$ form a complete set of real orthogonal four-vectors. These new spin wave functions satisfy the same properties as in eqs. (16.6) and (16.7) and are also defined off-shell, where they satisfy

$$\Lambda_{\mu\nu}(k) \equiv \sum_\lambda \varepsilon_\mu^{(\lambda)}(k)\varepsilon_\nu^{(\lambda)}(k)^* = -\left(g_{\mu\nu} - \frac{k_\mu k_\nu}{k^2}\right). \qquad (16.10)$$

Sums over λ will, of course, run over the set $\{0, +, -\}$ in this case. We leave it as an exercise to verify that rotations over an angle α around the axis pointing in the direction of $\vec{k}$ leaves $\varepsilon_\mu^{(0)}(k)$ invariant and transforms $\varepsilon_\mu^{(\pm)}(k)$ to $e^{\pm i\alpha}\varepsilon_\mu^{(\pm)}(k)$.

The Hamiltonian for the massive spin-one particles is given by

$$H = \int d_3\vec{k} \; \tfrac{1}{2}k_0(\vec{k}) \sum_\lambda \left(a_\lambda^\dagger(\vec{k})a_\lambda(\vec{k}) + a_\lambda(\vec{k})a_\lambda^\dagger(\vec{k})\right), \qquad (16.11)$$

which can be expressed in terms of three scalar fields φ_λ, $\lambda = 1, 2, 3$, as

$$
\begin{aligned}
H &= \int d_3\vec{k} \left(\tfrac{1}{2}|\tilde{\pi}_\lambda(\vec{k})|^2 + \tfrac{1}{2}(\vec{k}^2 + m^2)|\tilde{\varphi}_\lambda(\vec{k})|^2\right) \\
&= \int d_3\vec{x} \left(\tfrac{1}{2}\pi_\lambda^2(\vec{x}) + \tfrac{1}{2}(\partial_i\varphi_\lambda(\vec{x}))^2 + \tfrac{1}{2}m^2\varphi_\lambda^2(\vec{x})\right),
\end{aligned} \qquad (16.12)
$$

where we defined, as in eqs. (2.6) and (2.7),

$$\tilde{\varphi}_\lambda(\vec{k}) = (a_\lambda(\vec{k}) + a_\lambda^\dagger(-\vec{k}))/\sqrt{2k_0(\vec{k})},$$
$$\tilde{\pi}_\lambda(\vec{k}) = \frac{i}{2}\sqrt{2k_0(\vec{k})}(a_\lambda^\dagger(\vec{k}) - a_\lambda(-\vec{k})). \tag{16.13}$$

The corresponding Lagrangian would be given by

$$\mathcal{L}_\varphi = \sum_\lambda \left(\tfrac{1}{2}(\partial_\mu \varphi_\lambda(x))^2 - \tfrac{1}{2}m^2\varphi_\lambda^2(x) - \varphi_\lambda(x)J^{(\lambda)}(x)\right), \tag{16.14}$$

where we have added a source for each scalar field. We introduce also a field σ for the component of the vector field along k_μ. Writing

$$\tilde{A}_\mu(k) \equiv \frac{1}{m}\tilde{\sigma}(k)k_\mu + \sum_\lambda \tilde{\varphi}_\lambda(k)\varepsilon_\mu^{(\lambda)}(k),$$
$$\tilde{J}_\mu(k) \equiv \frac{1}{m}\tilde{j}(k)k_\mu + \sum_\lambda \tilde{J}^{(\lambda)}(k)\varepsilon_\mu^{(\lambda)}(k), \tag{16.15}$$

a simple calculation shows that $\mathcal{L}_A \equiv \mathcal{L}_\varphi - \frac{1}{2}[\partial_\mu\sigma(x)]^2 - \sigma(x)j(x)$. We see that σ decouples from the other components and behaves like a scalar particle with the *wrong* sign for the kinetic term. This would lead to serious inconsistencies, which are circumvented if we take $\partial_\mu J^\mu(x) = 0$, such that we can put $\sigma \equiv 0$.

It is important to realise that it is the Lorentz invariance that requires us to describe a spin-one particle by a four-vector. From the point of view of the scalar degrees of freedom, φ_λ, this invariance seems to be lost when we introduce interactions in the Lagrangian of eq. (16.14). Nevertheless, if we treat λ as a three-vector index (in some *internal* space) and demand the interactions to be O(3) invariant with respect to this index (i.e., invariance under rigid rotations *and* reflections in the internal space), then we claim that the resulting interactions do respect the Lorentz invariance. The reason is simple, because the O(3) invariance requires that the λ index is always pairwise contracted. Equations (16.10) and (16.15) guarantee that such a pair, written in terms of the vector field $A_\mu(x)$, is a Lorentz scalar as far as it concerns the dependence on A, which is sufficient if the Lagrangian is Lorentz invariant when treating λ as a dummy label. To eliminate the σ field we have to enforce $\partial_\mu A^\mu(x) = 0$, not only on-shell but also off-shell. This can be achieved by adding to eq. (16.2) a term $-\lambda(x)\partial_\mu A^\mu(x)$ (compare this to Problem 8). Using

$$\int \mathcal{D}A_\mu(x)\mathcal{D}\lambda(x) \exp\left[-i\int d_4x\, \lambda(x)\partial_\mu A^\mu(x)\right] = \int \mathcal{D}\tilde{A}_\mu(k) \prod_k \delta(k_\mu \tilde{A}^\mu(k)), \tag{16.16}$$

we see that the so-called Lagrange multiplier field $\lambda(x)$ plays the role of removing the unwanted degree of freedom. Since we modified the theory off-shell, the propagator in eq. (16.8) has to be changed also, by replacing $\Lambda(\vec{k})$ by its off-shell value $\Lambda(k)$ [eq. (16.10)]. If none of the vertices or sources couple to the σ field, we might just as well replace it by $-g_{\mu\nu}$. There are a number of other ways to eliminate the σ degree of freedom; see, e.g., Section 3-2-3 in Itzykson and Zuber. We will come back to massive vector particles in Chapter 19.

For the computation of the scattering matrix, we express the annihilation and creation operators in terms of the vector fields at $t = 0$, using the relations

$$a_\lambda(\vec{k}) = -\sqrt{2k_0(\vec{k})}\varepsilon_\mu^{(\lambda)}(\vec{k})^* \int \frac{d_3\vec{x}}{\sqrt{(2\pi)^3}}\, \hat{A}^\mu(\vec{x})e^{-i\vec{k}\cdot\vec{x}},$$

$$a_\lambda^\dagger(\vec{k}) = -\sqrt{2k_0(\vec{k})}\varepsilon_\mu^{(\lambda)}(\vec{k}) \int \frac{d_3\vec{x}}{\sqrt{(2\pi)^3}}\, \hat{A}^\mu(\vec{x})e^{i\vec{k}\cdot\vec{x}},$$

(16.17)

which in the path integral turn into

$$\hat{a}_\lambda^\dagger(\vec{k}) = \sqrt{2k_0(\vec{k})}\varepsilon_\mu^{(\lambda)}(\vec{k})\frac{-i\delta}{\delta \tilde{J}_\mu(\vec{k}, t = T_{\text{in}})},$$

$$\hat{a}_\lambda(\vec{k}) = \sqrt{2k_0(\vec{k})}\varepsilon_\mu^{(\lambda)}(\vec{k})^*\frac{-i\delta}{\delta \tilde{J}_\mu(-\vec{k}, t = T_{\text{out}})},$$

(16.18)

which is *identical* to eq. (9.3), when re-expressed in terms of $J^{(\lambda)}$. Like for scalar and fermion fields, there will be a mass and wave-function (denoted by Z_A) renormalisation, determined through the self-energy of the vector field, which is now a Lorentz tensor of rank two. It is proportional to $\Lambda_{\mu\nu}(k)$ [to guarantee that the scalar field σ introduced above decouples from the other fields; alternatively it can be seen as a consequence of the O(3) invariance with respect to the index λ]. We can consequently define

$$\Sigma_{\mu\nu}(p) \equiv \Lambda_{\mu\nu}(p)\Sigma_A(p).$$

(16.19)

The n-point Green's functions can now be written in terms of amputated Green's functions that carry four-vector indices for each external spin-one line,

$$G_c^{(n)}(J_\mu) \equiv \int \left\{ \prod_{j=1}^n d_4 p_j\, \tilde{J}^{\mu_j}(p_j)\frac{-i\Lambda_{\mu_j\nu_j}(p_j)}{p_j^2 - m^2 - \Sigma_A(p_j) + i\varepsilon} \right\}$$

$$\times G_c^{\text{amp}}(p_1, p_2, \ldots, p_n)^{\nu_1, \nu_2, \cdots, \nu_n}.$$

(16.20)

Using the fact that on-shell $\Lambda_\mu{}^\nu(\vec{k})\varepsilon_\nu^{(\lambda)}(\vec{k}) = -\varepsilon_\mu^{(\lambda)}(\vec{k})$, we find for the *incoming* spin-one line a wave-function factor $\sqrt{Z_A}\varepsilon_\mu^{(\lambda)}(\vec{k})$ and for the *outgoing* line a factor $\sqrt{Z_A}\varepsilon_\mu^{(\lambda)}(\vec{k})^*$.

For a massless spin-one field (the photon), we would expect the helicity zero component of the vector field to be absent. First we have to redefine, however, what we would mean by the zero helicity component, because eq. (16.9) is singular in the limit of zero mass. We take as our definition

$$\varepsilon_\mu^{(0)}(k) = n_\mu(k) \equiv \tfrac{1}{2}\sqrt{2}\left(1, \frac{\vec{k}}{|\vec{k}|}\right), \quad \varepsilon_\mu^{(\pm)}(k) = \left(0, \vec{s}_\pm(\vec{k})\right), \qquad (16.21)$$

with $\vec{k} \cdot \vec{s}_\pm(\vec{k}) = 0$. These still form with k_μ four independent four-vectors and $\varepsilon_\mu^{(\pm)}(\vec{k})$ are still transverse polarisations. However, it is no longer true that $k^\mu \varepsilon_\mu^{(0)}(\vec{k})$ will vanish on-shell (i.e., at $k^2 = 0$). This is easily seen to imply that *on-shell* $a_0(\vec{k}) = 0$. In other words, on-shell there is no longitudinal component for the photon. The extra degree of freedom is removed by the gauge invariance of the massless vector field, as was discussed at the end of Chapter 4 and in Problem 9. To perform the quantisation of the theory, one can go about as in the massive case. Due to the presence of the four-vector $n(k)$, it will be much more cumbersome to demonstrate the Lorentz invariance. In Chapter 20 it will be shown how in principle in any gauge the path integral can be defined and that the result is independent of the chosen gauge. One could then choose a gauge that allows us to show the equivalence between the path integral and the canonical quantisation. However, it is the great advantage of the path integral formulation that calculations can be performed in a gauge in which the Lorentz invariance is manifest. The gauge most suitable for that purpose is, of course, the Lorentz gauge $\partial_\mu A^\mu(x) = 0$; see eqs. (4.21) and (4.22). The propagator is read off from eq. (4.26)

$$\frac{-\left(g_{\mu\nu} - (1-\tfrac{1}{\alpha})\frac{k_\mu k_\nu}{k^2+i\varepsilon}\right)}{k^2 + i\varepsilon} \equiv \frac{\Lambda_{\mu\nu}^{(\alpha)}(k)}{k^2 + i\varepsilon}, \qquad (16.22)$$

where α is an arbitrary parameter, on which physical observables like cross sections and decay rates should not depend. It is in general *not* true anymore that the self-energy is proportional to $\Lambda_{\mu\nu}^{(\alpha)}(k)$, but the gauge invariance does guarantee that $\Sigma_{\mu\nu}(k)\tilde{j}^\nu(k)$ is independent of α for any conserved current; i.e., $k_\mu \tilde{j}^\mu(k) = 0$. It can be shown that this in general implies

$$\Sigma_{\mu\nu}(k) = \Lambda_{\mu\nu}^{(\alpha')}(k)\Sigma(k), \qquad (16.23)$$

for some, possibly infinite, α'. Apart from a wave-function renormalisation (Z_A), the gauge-fixing parameter α will in principle have to be renormalised too. One still has for any value of α that $\Lambda^{(\alpha)}(k)_\mu{}^\nu \varepsilon_\nu^{(\pm)}(\vec{k}) = -\varepsilon_\mu^{(\pm)}(\vec{k})$. The wave-function factors for external photon lines are therefore identical to the ones for the massive case, except that now only two helicity states can appear. It is an important consequence of gauge invariance that unphysical degrees of freedom decouple in the physical amplitudes. It also implies that the

TABLE 16.1

Feynman rules for photons.

$\equiv \dfrac{-\left(g_{\mu\nu} - (1 - \dfrac{1}{\alpha})\dfrac{k_\mu k_\nu}{k^2 + i\varepsilon}\right)}{k^2 + i\varepsilon}$		photon propagator (Lorentz gauge)
$\equiv \sqrt{Z_A}\varepsilon_\mu^{(\lambda)}(\vec{k})$		incoming photon
$\equiv \sqrt{Z_A}\varepsilon_\mu^{(\lambda)}(\vec{k})^*$		outgoing photon

self-energy vanishes on shell (see Problem 39), such that it will *not* give rise to a renormalisation of the mass. The photon remains massless. That the gauge invariance must be crucial here is clear, as a massive photon would have one extra degree of freedom. We will come back to this point in Chapter 19. In Table 16.1 we summarise the Feynman rules.

17

Quantum Electrodynamics—QED

DOI: 10.1201/b15364-17

QED is the field theory that describes the interaction between the photon and the charged fermions. In the Lorentz gauge [see Equations (4.21) and (4.22)], the Lagrangian is given by

$$\mathcal{L} = -\tfrac{1}{4}F_{\mu\nu}(x)F^{\mu\nu}(x) - \tfrac{1}{2}\alpha\left(\partial_\mu A^\mu(x)\right)^2 + \sum_f \overline{\Psi}_f(x)(i\gamma^\mu D_\mu - m_f)\Psi_f(x).$$

(17.1)

Here f is the so-called flavour index, which distinguishes the various types of fermions (electrons, protons, etc.). The covariant derivative D_μ is given as before [see Equations (3.35) and (14.11)] by

$$D_\mu \Psi_f(x) = \left(\partial_\mu + iq_f A_\mu(x)\right)\Psi_f(x).$$

(17.2)

For electrons we have $q_f = -e$ and for protons $q_f = e$. For $\alpha = 0$ the Lagrangian is invariant under gauge transformations

$$A_\mu(x) \to A_\mu(x) + \partial_\mu \Lambda(x), \qquad \Psi_f(x) \to \exp\left(-iq_f \Lambda(x)\right)\Psi_f(x).$$

(17.3)

The Feynman rules are collected in Table 17.1.

Before calculating cross sections we wish to discuss in more detail the helicity of the fermions and its relation to charge conjugation C. The latter relates, say, electrons to positrons, or in general particles to antiparticles, which is an important symmetry of the theory. It, as well as parity (P) and time reversal (T) symmetry, can be separately broken, but the combination **CPT** is to be unbroken to allow for a local, relativistic invariant field theory. The spin components of the solutions in Equation (13.8) were based on a decomposition along the z axis in the rest frame. Helicity, as for the photon, is defined by decomposing the spin in the direction of motion, $\vec{k}$. It is hence defined in terms of the eigenvalues of the operator

$$\hat{k} \cdot \vec{J} \equiv \frac{k_i}{4|\vec{k}|}\varepsilon_{ijk}\sigma_{jk} = \begin{pmatrix} \tfrac{1}{2}\hat{k}\cdot\vec{\sigma} & \oslash \\ \oslash & \tfrac{1}{2}\hat{k}\cdot\vec{\sigma} \end{pmatrix}.$$

(17.4)

($\vec{J}$ is the spin part of the angular momentum operator, the equivalent of $\tfrac{1}{2}\vec{\sigma}$ for a two-component spinor.) This holds both in the Dirac and Weyl

TABLE 17.1

Feynman rules for QED.

$\underset{\mu \qquad \nu}{\overset{k}{\wwwww}}$ $\equiv$	$\dfrac{-\left(g_{\mu\nu} - (1 - \dfrac{1}{\alpha})\dfrac{k_\mu k_\nu}{k^2 + i\varepsilon}\right)}{k^2 + i\varepsilon}$	photon propagator (Lorentz gauge)
$\underset{k_1 \qquad k_2}{\overset{\mu \; k_3}{b \; \wedge \; a}}$ $\equiv$	$q\gamma^\mu_{ab}$ and $k_1 = k_2 + k_3$	photon vertex (fermion charge is q)
$\overrightarrow{b \; k \; a}$ $\equiv$	$\left[\dfrac{1}{\not{k} - m + i\varepsilon}\right]_{ab}$	fermion propagator

representations. It is easy to verify that $[\hat{k} \cdot \vec{J}, \not{k}] = 0$, e.g., by making use of the fact that in the Dirac representation

$$\not{k} = \begin{pmatrix} k_0 1_2 & \vec{k} \cdot \vec{\sigma} \\ -\vec{k} \cdot \vec{\sigma} & -k_0 1_2 \end{pmatrix}. \tag{17.5}$$

This implies that we can choose $u_0^{(\alpha)}$ and $v_0^{(\alpha)}$ to be eigenstates of the helicity operator $\hat{k} \cdot \vec{J}$ (consequently they become functions of $\hat{k}$)

$$\hat{k} \cdot \vec{J} \; \tilde{\Psi}(\vec{k}) = \pm \tfrac{1}{2} \tilde{\Psi}(\vec{k}). \tag{17.6}$$

Instead of the label α, we can use $\pm$ to indicate the helicity and we have

$$u^\pm(\vec{k}) = \frac{(\not{k} + m)}{\sqrt{m + |k_0|}} u_0^\pm(\hat{k}), \quad \hat{k} \cdot \vec{J} u_0^\pm(\hat{k}) = \pm \tfrac{1}{2} u_0^\pm(\hat{k}),$$

$$v^\pm(\vec{k}) = \frac{(-\not{k} + m)}{\sqrt{m + |k_0|}} v_0^\pm(\hat{k}), \quad \hat{k} \cdot \vec{J} v_0^\pm(\hat{k}) = \mp \tfrac{1}{2} v_0^\pm(\hat{k}). \tag{17.7}$$

Note the flip of helicity for the positron wave functions. For $\vec{k} = (0, 0, k)$ these eigenstates coincide with the decomposition in Equation (13.8). It is clear that we can define

$$u_0^\pm(\hat{k}) = \begin{pmatrix} \varphi_\pm(\hat{k}) \\ 0 \\ 0 \end{pmatrix}, \quad \hat{k} \cdot \vec{\sigma} \, \varphi_\pm(\hat{k}) = \pm \varphi_\pm(\hat{k}),$$

$$v_0^\pm(\hat{k}) = \begin{pmatrix} 0 \\ 0 \\ \chi_\pm(\hat{k}) \end{pmatrix}, \quad \hat{k} \cdot \vec{\sigma} \, \chi_\pm(\hat{k}) = \mp \chi_\pm(\hat{k}), \tag{17.8}$$

with $\varphi_\pm$ and $\chi_\pm$ each an orthonormal set of two-component spinors. They can be related to each other by

$$\chi_\pm(\hat{k}) \equiv -i\sigma_2 \varphi_\pm^*(\hat{k}). \tag{17.9}$$

Indeed, when we use that

$$\sigma_2 \sigma_i \sigma_2 = -\sigma_i^*, \qquad i = 1, 2, 3, \tag{17.10}$$

which expresses the fact that SU(2) is so-called pseudo real, we find

$$\hat{k} \cdot \vec{\sigma} \left(-i\sigma_2 \, \varphi_\pm^*(\hat{k})\right) = \left(-i\hat{k} \cdot \vec{\sigma}^* \sigma_2 \varphi_\pm(\hat{k})\right)^* = \left(i\sigma_2 \, \hat{k} \cdot \vec{\sigma} \, \varphi_\pm(\hat{k})\right)^*$$
$$= \mp \left(-i\sigma_2 \, \varphi_\pm^*(\hat{k})\right). \tag{17.11}$$

As Equation (17.9) relates the components of the electron wave function to those of the positron wave function, it is the basis of the charge conjugation symmetry, which relates the solutions of the Dirac equation to solutions of the complex conjugate Dirac equation [see Equation (12.31)], which indeed interchanges positive and negative energy solutions, i.e., particles and antiparticles. To formulate this symmetry in the four-component spinor language, one introduces the charge conjugation matrix (in the Dirac representation)

$$C \equiv -i\gamma^0\gamma^2 = \begin{pmatrix} \oslash & -i\sigma_2 \\ -i\sigma_2 & \oslash \end{pmatrix}, \tag{17.12}$$

which satisfies

$$C^{-1} = C^\dagger = -C, \qquad C\gamma_\mu C^{-1} = -\gamma_\mu^t. \tag{17.13}$$

This can be proven from the explicit form of the Dirac matrices. The equivalent of Equation (17.10) is given by

$$\gamma^2 \gamma_\mu \gamma_2 = -\gamma_\mu^*. \tag{17.14}$$

It is now easy to verify that

$$v_\pm(\vec{k}) = C\bar{u}_\pm^t(\vec{k}), \qquad u_\pm(\vec{k}) = C\bar{v}_\pm^t(\vec{k}). \tag{17.15}$$

We just need to prove one of these identities, because charge conjugation is an involution, i.e., applying it twice gives the identity

$$C\overline{\left(C\overline{\Psi}^t\right)}^t = C\gamma_0^t C^* \gamma_0^\dagger \Psi = \Psi. \tag{17.16}$$

We find

$$C\bar{u}_\pm^t(\vec{k}) = C\gamma_0^t u_\pm^*(\vec{k}) = i\gamma^2 u_\pm^*(\vec{k}) = i\gamma^2 \frac{(\not{k}^* + m)}{\sqrt{m + |k_0|}} \begin{pmatrix} \varphi_\pm^*(\hat{k}) \\ 0 \\ 0 \end{pmatrix}$$

$$= i \frac{(-\not{k} + m)}{\sqrt{m + |k_0|}} \gamma^2 \begin{pmatrix} \varphi_\pm^*(\hat{k}) \\ 0 \\ 0 \end{pmatrix} = \frac{(-\not{k} + m)}{\sqrt{m + |k_0|}} \begin{pmatrix} \oslash & i\sigma_2 \\ -i\sigma_2 & \oslash \end{pmatrix} \begin{pmatrix} \varphi_\pm^*(\hat{k}) \\ 0 \\ 0 \end{pmatrix}$$

$$= \frac{(-\not{k} + m)}{\sqrt{m + |k_0|}} \begin{pmatrix} 0 \\ 0 \\ \chi_\pm(\hat{k}) \end{pmatrix} = v_\pm(\vec{k}). \tag{17.17}$$

Under charge conjugation the charge that appears in the covariant derivative in Equation (17.2) should change sign too. To show this we multiply the complex conjugate of the Dirac equation with $i\gamma^2$

$$i\gamma^2 \left[(-i\gamma_\mu(\partial^\mu - ieA^\mu) + m) \Psi \right]^* = (i\gamma^2\gamma_\mu^*\gamma_2(\partial^\mu + ieA^\mu) + m)(i\gamma^2\Psi^*)$$

$$= (-i\gamma_\mu(\partial^\mu + ieA^\mu) + m)(C\overline{\Psi}^t). \tag{17.18}$$

That charge conjugation is really a symmetry of the quantum theory is most convincingly demonstrated by the fact that the Dirac Lagrangian is invariant under charge conjugation. Using $\overline{C\overline{\Psi}^t} = \Psi^t\gamma_0^*C^\dagger\gamma_0 = -\Psi^tC^{-1}$, the anticommuting properties of the fermi fields and partial integration, we find

$$\int d_4 x \, \overline{C\overline{\Psi}^t} \left(i\gamma_\mu(\partial^\mu + ieA^\mu) - m \right) C\overline{\Psi}^t$$

$$= \int d_4 x \, -\Psi^t C^{-1} \left(i\gamma_\mu(\partial^\mu + ieA^\mu) - m \right) C\overline{\Psi}^t$$

$$= \int d_4 x \, -\Psi^t \left(-i\gamma_\mu^t(\partial^\mu + ieA^\mu) - m \right) \overline{\Psi}^t$$

$$= \int d_4 x \, \overline{\Psi} \left(i\gamma_\mu(\partial^\mu - ieA^\mu) - m \right) \Psi. \tag{17.19}$$

In particular we see that the electromagnetic current generated by the fermi fields transforms as required for the interchange of particles and antiparticles, under which the charge changes sign

$$j_\mu = \overline{\Psi}\gamma_\mu\Psi \overset{C}{\rightarrow} -\overline{\Psi}\gamma_\mu\Psi. \tag{17.20}$$

An important consequence of the charge conjugation symmetry is Furry's theorem, which states that a fermionic loop with an odd number of vertices

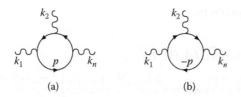

(a)   (b)

FIGURE 17.1
Fermion loops with an odd number of photons.

will not contribute to the amplitude. Consider a fermion loop as in Figure 17.1(a) for which the Feynman rules lead to the expression (note that spinor index contractions run against the arrow of the fermion line and $\sum_i k_i = 0$)

$$q^n \text{Tr}\left(\frac{1}{\not{p}-m+i\varepsilon}\gamma^{\mu_1}\frac{1}{\not{p}+\not{k}_1-m+i\varepsilon}\gamma^{\mu_2}\frac{1}{\not{p}+\not{k}_1+\not{k}_2-m+i\varepsilon}\gamma^{\mu_3}\cdots\gamma^{\mu_n}\right).$$
(17.21)

Using the fact that for any matrix A we have $\text{Tr}(A) = \text{Tr}(A^t) = \text{Tr}(C\,A^t C^{-1})$, we convert Equation (17.21) to the expression

$$(-q)^n \text{Tr}\left(\gamma^{\mu_n}\cdots\gamma^{\mu_3}\frac{1}{-\not{p}-\not{k}_1-\not{k}_2-m+i\varepsilon}\gamma^{\mu_2}\frac{1}{-\not{p}-\not{k}_1-m+i\varepsilon}\gamma^{\mu_1}\right.$$
$$\left.\times\frac{1}{-\not{p}-m+i\varepsilon}\right),$$
(17.22)

which is exactly $(-1)^n$ times the result of the Feynman diagram that is obtained by inverting the orientation of the fermion line (i.e., the vertices are connected in the reversed order) as indicated in Figure 17.1(*b*). As both diagrams will occur, their contributions will cancel whenever n is odd. It confirms the intuition that particles and antiparticles contribute equally, except for their opposite charge factors $(\pm q)^n$.

We will now calculate the cross section for electron–electron scattering (the so-called Møller cross section). In lowest nontrivial order there are only two diagrams that contribute, as indicated in Figure 17.2.

The labels t_i and s_i indicate the helicities of the incoming and outgoing electrons. The scattering matrix (ignoring the time-dependent phase factor)

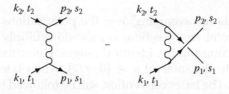

FIGURE 17.2
Diagrams that contribute to the electron–electron scattering.

for this process is given by

$$
_{\text{out}}< (\vec{p}_1, s_1), (\vec{p}_2, s_2)|(\vec{k}_1, t_1), (\vec{k}_2, t_2) >_{\text{in}}
$$

$$
= -i(2\pi)^4 \frac{\delta_4(p_1 + p_2 - k_1 - k_2)\mathcal{M}(\{(-p_1, s_1), (-p_2, s_2)\}, \{(k_1, t_1), (k_2, t_2)\})}{\sqrt{2p_0^{(1)}(\vec{p}_1)(2\pi)^3 2p_0^{(2)}(\vec{p}_2)(2\pi)^3 2k_0^{(1)}(\vec{k}_1)(2\pi)^3 2k_0^{(2)}(\vec{k}_2)(2\pi)^3}}
$$

$$
= \frac{i\delta_4(p_1 + p_2 - k_1 - k_2)}{(4\pi)^2 \sqrt{p_0^{(1)} p_0^{(2)} k_0^{(1)} k_0^{(2)}}} \left\{ \frac{\bar{u}_{s_1}(p_1)e\gamma^\mu u_{t_1}(k_1)g_{\mu\nu}\bar{u}_{s_2}(p_2)e\gamma^\nu u_{t_2}(k_2)}{(k_1 - p_1)^2 + i\varepsilon} \right.
$$

$$
\left. - \frac{\bar{u}_{s_1}(p_1)e\gamma^\mu u_{t_2}(k_2)g_{\mu\nu}\bar{u}_{s_2}(p_2)e\gamma^\nu u_{t_1}(k_1)}{(k_1 - p_2)^2 + i\varepsilon} \right\}. \tag{17.23}
$$

The relative minus sign is, of course, a consequence of the so-called Fermi–Dirac statistics, which implements the Pauli principle. We got rid of the gauge-dependent part of the photon propagator (see Table 17.1) by using the fact that the currents generated by $\bar{u}_s(p)\gamma^\mu u_t(k)$ are conserved, such that

$$
\bar{u}_s(p)\gamma^\mu u_t(k)(p_\mu - k_\mu) = \bar{u}_s(p)[(\not{p} - m) - (\not{k} - m)]u_t(k) = 0, \tag{17.24}
$$

because on-shell $(\not{k} - m)u(k) = 0$ [and hence also $\bar{u}(k)(\not{k} - m) = 0$]. Indirectly, through current conservation, this is related to gauge invariance. It guarantees that the longitudinal component of the photon does not contribute to the scattering matrix, which is thus seen not to depend on the gauge-fixing parameter α.

The differential cross section for *unpolarised* electron–electron scattering is given by [see Equation (10.12); from now on we will drop the distinction between $\vec{k}_i$ and k_i]

$$
d\sigma = \sum_{s_1, s_2} \frac{d_3\vec{p}_1}{2p_0(\vec{p}_1)(2\pi)^3} \frac{d_3\vec{p}_2}{2p_0(\vec{p}_2)(2\pi)^3} \frac{(2\pi)^4\delta_4(p_1 + p_2 - k_1 - k_2)}{4\sqrt{(k_1 \cdot k_2)^2 - m^4}}
$$

$$
\times \frac{1}{4} \sum_{t_1, t_2} |\mathcal{M}(\{(-p_1, s_1), (-p_2, s_2)\}, \{(k_1, t_1), (k_2, t_2)\})|^2, \tag{17.25}
$$

where $\frac{1}{4}\sum_{t_1, t_2}$ stands for averaging over the polarisations of the incoming electrons. For the total cross section, we should multiply with a factor of $\frac{1}{2}$ to avoid double counting the identical outgoing electrons, or restrict the scattering angle θ to the interval $\theta \in [0, \pi/2]$, when integrating over the outgoing momenta. The latter convention will be followed here. In the center of mass system, the scattered particles move back to back in a direction which is only determined modulo π, which is why $\theta \in [0, \pi/2]$, with θ measured from the incoming particle direction (also defined modulo π).

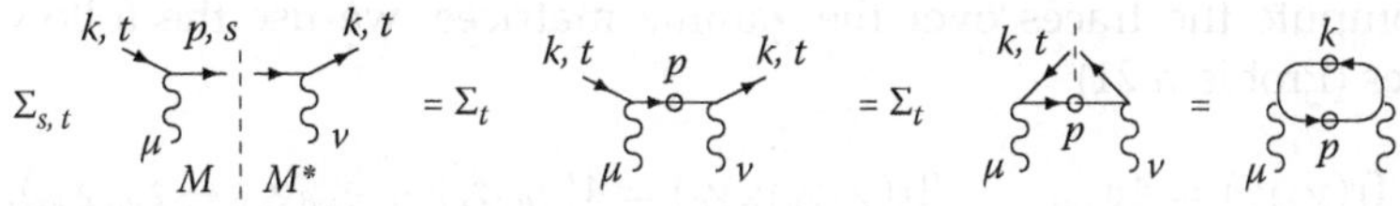

FIGURE 17.3
Graphical representation of Equation (17.26).

To calculate $|\mathcal{M}|^2$ we use

$$\sum_{s,t} \bar{u}_s(p)\gamma_\mu u_t(k)\big(\bar{u}_s(p)\gamma_\nu u_t(k)\big)^* = \sum_{s,t} \bar{u}_s(p)\gamma_\mu u_t(k)\,\bar{u}_t(k)\gamma_0\gamma_\nu^\dagger\gamma_0 u_s(p)$$

$$= \mathrm{Tr}\Big(\gamma_\mu \sum_t u_t(k)\otimes\bar{u}_t(k)\gamma_\nu \sum_s u_s(p)\otimes\bar{u}_s(p)\Big) = \mathrm{Tr}\big(\gamma_\mu(\slashed{k}+m)\gamma_\nu(\slashed{p}+m)\big),$$

$$(17.26)$$

which can be represented graphically as in Figure 17.3. Hence, we add a Feynman rule for the so-called *cut* fermion propagator

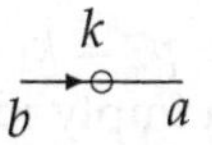

$$\underset{b}{\overset{k}{\longrightarrow\!\!\circ\!\!\longrightarrow}}\underset{a}{} = \mathrm{sign}(k_0)(\slashed{k}+m)_{ab}.$$

$$(17.27)$$

For antiparticles $k_0 < 0$ [see Equation (13.18) for the extra minus sign]. These results can be generalised to other fields too, by noting that our conventions have been such that the propagators can be written as

$$\frac{\sum_\beta \phi_\beta(\vec{k})\otimes\bar{\phi}_\beta(\vec{k})^*}{k^2 - m^2 + i\varepsilon},$$

$$(17.28)$$

where $\phi_\beta(\vec{k})$ are the wave functions for the incoming lines and $\bar{\phi}_\beta(\vec{k})^*$ for the outgoing lines, with β labelling the internal degrees of freedom [compare this to Equation (16.8)].

We can now use this result to compute $|\mathcal{M}|^2$

$$\sum_{s_1,s_2,t_1,t_2} |\mathcal{M}|^2 = \left(\;\;\;\;\;\;\;\;\; - \;\;\;\;\;\;\;\;\;\right) + (p_1 \longleftrightarrow p_2)$$

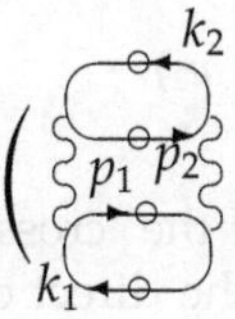

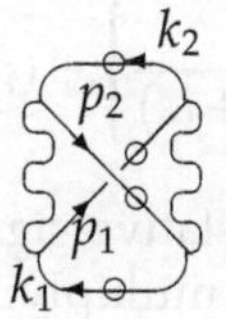

$$= e^4 \Bigg\{ \frac{\mathrm{Tr}\big[\gamma^\mu(\slashed{k}_1+m)\gamma^\nu(\slashed{p}_1+m)\big]\,\mathrm{Tr}\big[\gamma_\mu(\slashed{k}_2+m)\gamma_\nu(\slashed{p}_2+m)\big]}{\big((k_1-p_1)^2+i\varepsilon\big)\big((k_2-p_2)^2+i\varepsilon\big)}$$

$$- \frac{\mathrm{Tr}\big[\gamma^\mu(\slashed{k}_1+m)\gamma^\nu(\slashed{p}_2+m)\gamma_\mu(\slashed{k}_2+m)\gamma_\nu(\slashed{p}_1+m)\big]}{\big((k_1-p_1)^2+i\varepsilon\big)\big((k_1-p_2)^2+i\varepsilon\big)}\Bigg\}$$

$$+(p_1 \longleftrightarrow p_2).$$

$$(17.29)$$

To compute the traces over the gamma matrices, we use the following identities (Problem 21):

$$\text{Tr}(\gamma_\mu \gamma_\nu) = 4g_{\mu\nu}, \qquad \text{Tr}(\gamma_\mu \gamma_\nu \gamma_\alpha \gamma_\beta) = 4(g_{\mu\nu}g_{\alpha\beta} + g_{\mu\beta}g_{\alpha\nu} - g_{\mu\alpha}g_{\nu\beta}),$$

$$\sum_\mu \gamma_\mu \gamma_\alpha \gamma_\beta \gamma^\mu = 4g_{\alpha\beta}, \qquad \sum_\mu \gamma_\mu \gamma_\nu \gamma_\alpha \gamma_\beta \gamma^\mu = -2\gamma_\beta \gamma_\alpha \gamma_\nu, \qquad (17.30)$$

and the fact that the trace over an odd number of gamma matrices vanishes. This implies

$$\text{Tr}(\gamma_\mu (\slashed{k} + m)\gamma_\nu (\slashed{p} + m)) = \text{Tr}(\gamma_\mu \slashed{k} \gamma_\nu \slashed{p}) + 4m^2 g_{\mu\nu}$$
$$= 4(m^2 - k \cdot p)g_{\mu\nu} + 4k_\mu p_\nu + 4k_\nu p_\mu, \qquad (17.31)$$

and

$$\sum_\mu \gamma_\mu (\slashed{k} + m)\gamma_\nu (\slashed{p} + m)\gamma^\mu = -2\slashed{p}\gamma_\nu \slashed{k} - 2m^2 \gamma_\nu + 4mk_\nu + 4mp_\nu. \quad (17.32)$$

Together with momentum conservation ($p_1 + p_2 = k_1 + k_2$) and the on-shell conditions ($p_1^2 = p_2^2 = k_1^2 = k_2^2 = m^2$), which imply identities like $p_1 \cdot p_2 = k_1 \cdot k_2$, we find

$$\sum_{s_1,s_2,t_1,t_2} |\mathcal{M}|^2$$

$$= e^4 \left\{ \frac{16(g^{\mu\nu}(m^2 - k_1 \cdot p_1) + k_1^\mu p_1^\nu + k_1^\nu p_1^\mu)(g_{\mu\nu}(m^2 - k_2 \cdot p_2) + k_2^2 p_v^2 + k_v^2 p_\mu^2)}{((k_1 - p_1)^2 + i\varepsilon)((k_2 - p_2)^2 + i\varepsilon)} \right.$$

$$\left. -\kappa \frac{\text{Tr}[(2\slashed{p}_2 \gamma^\nu \slashed{k}_1 + 2m^2\gamma^\nu - 4mk_1^\nu - 4mp_2^\nu)(\slashed{k}_2 + m)\gamma_\nu(\slashed{p}_1 + m)]}{((k_1 - p_1)^2 + i\varepsilon)((k_1 - p_2)^2 + i\varepsilon)} \right\} + (p_1 \leftrightarrow p_2)$$

$$= 32e^4 \left\{ \frac{(k_1 \cdot k_2)^2 + (k_1 \cdot p_2)^2 + 2m^2(k_1 \cdot p_2 - k_1 \cdot k_2)}{((k_1 - p_1)^2 + i\varepsilon)^2} \right.$$

$$\left. -\kappa \frac{(k_1 \cdot k_2)^2 - 2m^2 k_1 \cdot k_2}{((k_1 - p_1)^2 + i\varepsilon)((k_1 - p_2)^2 + i\varepsilon)} \right\} + (p_1 \leftrightarrow p_2). \qquad (17.33)$$

The parameter κ determines the relative sign of the 'crossed' diagrams in Equation (17.29), which arise from multiplying the direct electron–electron scattering diagram with the complex conjugate of the one where the outgoing fermion lines were crossed. For Fermi–Dirac statistics $\kappa \equiv -1$. By keeping track of the dependence on κ, one sees how scattering experiments can be used to verify the anticommuting nature of the electrons.

We finally perform some kinematics and express the differential cross section in terms of the scattering angle θ. We define in the center of mass frame

$$k_1^0 = k_2^0 = p_1^0 = p_2^0 \equiv E, \qquad \vec{p}_1 = -\vec{p}_2 \equiv \vec{p}, \vec{k}_1 = -\vec{k}_2 \equiv \vec{k} \quad \text{with} \quad |\vec{p}| = |\vec{k}|. \qquad (17.34)$$

Defining θ to be the angle between $\vec{k}$ and $\vec{p}$, i.e. $\vec{p} \cdot \vec{k} = \vec{k}^2 \cos \theta$, we have

$$
\begin{aligned}
k_1 \cdot k_2 &= E^2 + \vec{k}^2 = 2E^2 - m^2, \\
k_1 \cdot p_1 &= E^2 - \vec{k}^2 \cos \theta = E^2(1 - \cos \theta) + m^2 \cos \theta, \\
k_1 \cdot p_2 &= E^2 + \vec{k}^2 \cos \theta = E^2(1 + \cos \theta) - m^2 \cos \theta, \\
(p_2 - k_1)^2 &= -4\vec{k}^2 \cos^2(\tfrac{1}{2}\theta), \quad (p_1 - k_1)^2 = -4\vec{k}^2 \sin^2(\tfrac{1}{2}\theta), \\
(k_2 \cdot k_1)^2 - m^4 &= 4E^2(E^2 - m^2) = 4\vec{k}^2 E^2.
\end{aligned}
\tag{17.35}
$$

Finally we use the identity (Ω is the solid angle, $d\Omega = \sin\theta d\theta d\phi$)

$$
\int d_3\vec{p}_1 d_3\vec{p}_2\, \delta_4(p_1 + p_2 - k_1 - k_2) = \int p^2 dp d\Omega\, \delta\left(2\sqrt{\vec{p}^2 + m^2} - 2\sqrt{\vec{k}^2 + m^2}\right)
$$

$$
= \int d\Omega\, \tfrac{1}{2} E |\vec{k}|.
\tag{17.36}
$$

Collecting all terms we find for Equation (17.25) the result

$$
\frac{d\sigma}{d\Omega} = \frac{1}{(2E(2\pi)^3)^2} \cdot \frac{(2\pi)^4 \tfrac{1}{2} E |\vec{k}|}{4\sqrt{4\vec{k}^2 E^2}} \cdot \tfrac{1}{4} \sum_{s_1, s_2, t_1, t_2} |\mathcal{M}|^2 = \frac{1}{2^{10} \pi^2 E^2} \sum_{s_1, s_2, t_1, t_2} |\mathcal{M}|^2,
\tag{17.37}
$$

with $\sum_{s_1, s_2, t_1, t_2} |\mathcal{M}|^2$ given by

$$
32e^4 \left\{ \frac{(2E^2 - m^2)^2 + [E^2(1 + \cos\theta) - m^2 \cos\theta]^2 + 2m^2[E^2(1 + \cos\theta) - m^2\cos\theta + m^2 - 2E^2]}{16(\vec{k}^2)^2 \sin^4(\tfrac{1}{2}\theta)} \right.
$$
$$
+ \frac{(2E^2 - m^2)^2 + [E^2(1 - \cos\theta) + m^2 \cos\theta]^2 + 2m^2[E^2(1 - \cos\theta) + m^2\cos\theta + m^2 - 2E^2]}{16(\vec{k}^2)^2 \cos^4(\tfrac{1}{2}\theta)}
$$
$$
\left. - \kappa\, \frac{2(2E^2 - m^2)(2E^2 - 3m^2)}{16(\vec{k}^2)^2 \sin^2(\tfrac{1}{2}\theta) \cos^2(\tfrac{1}{2}\theta)} \right\}
$$

$$
= \frac{32e^4}{(\vec{k}^2)^2} \left\{ \frac{[(2E^2 - m^2)^2 + E^4 + (E^2 - m^2)^2 \cos^2\theta + 2m^2(m^2 - E^2)](\cos^4(\tfrac{1}{2}\theta) + \sin^4(\tfrac{1}{2}\theta))}{\sin^4\theta} \right.
$$
$$
+ \frac{\cos\theta[2E^2(E^2 - m^2) + 2m^2(E^2 - m^2)](\cos^4(\tfrac{1}{2}\theta) - \sin^4(\tfrac{1}{2}\theta))}{\sin^4\theta}
$$
$$
\left. - \kappa\, \frac{2(2E^2 - m^2)(2E^2 - 3m^2)}{4\sin^2\theta} \right\}
$$

$$
= \frac{16e^4}{(\vec{k}^2)^2} \left\{ (E^2 - m^2)^2 + \frac{4(2E^2 - m^2)^2}{\sin^4\theta} - \frac{3(2E^2 - m^2)^2 - m^4 + \kappa(2E^2 - m^2)(2E^2 - 3m^2)}{\sin^2\theta} \right\},
\tag{17.38}
$$

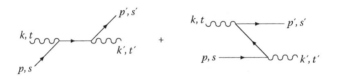

FIGURE 17.4
Diagrams that contribute to the electron–photon scattering.

yielding ($\alpha_e = \frac{e^2}{4\pi} = \frac{e^2}{4\pi\hbar c} \approx \frac{1}{137}$ is the fine-structure constant and $\kappa = -1$)

$$\frac{d\sigma}{d\Omega} = \frac{\alpha_e^2(2E^2 - m^2)^2}{4E^2(E^2 - m^2)^2}\left\{ \frac{4}{\sin^4\theta} - \frac{3}{\sin^2\theta} + \frac{(E^2 - m^2)^2}{(2E^2 - m^2)^2}\left(1 + \frac{4}{\sin^2\theta}\right)\right\}.$$

(17.39)

This cross section is invariant under $\theta \to \pi - \theta$, such that we cannot tell the two outgoing electrons apart, as it should be. For electron–electron scattering we have to put $\kappa = -1$, but we see from the dependence on κ in Equation (17.38) that one can easily distinguish experimentally if electrons behave according to the Fermi–Dirac statistics.

In Problem 29 electron–positron scattering is studied within the final state an electron and a positron (Bhabha scattering) or a muon and an antimuon. Both for $e^-e^- \longrightarrow e^-e^-$ and $e^-e^+ \longrightarrow e^-e^+$, one cannot take θ too close to zero (or π for e^-e^-). Apart from the fact that the detector would be in the way of the beam, it is fundamentally impossible to distinguish the scattered particles at $\theta = 0$ (and $\theta = \pi$ for e^-e^-) from those in the beam. The divergence of the differential cross section was therefore to be expected. For $e^-e^+ \longrightarrow \mu^-\mu^+$ this divergence is absent and one can define the total cross section by integrating over all angles. For $E \gg m_e$ and $E \gg m_\mu$ one finds (see Problem 29) $\sigma = \frac{1}{3}\pi\alpha_e^2\hbar^2c^2/E^2 = 21.7\,\text{nb}/E^2(\text{GeV})$.

We now discuss electron–photon scattering, also known as Compton scattering. The resulting cross section is called the Klein–Nishina formula. There are again two diagrams that contribute in lowest nontrivial order to the scattering matrix, Figure 17.4. The cross section is now given by

$$d\sigma = \sum_{s',t'} \frac{d_3\vec{p}\,'}{2p_0'(\vec{p}\,')(2\pi)^3} \frac{d_3\vec{k}'}{2k_0'(\vec{k}')(2\pi)^3} \frac{(2\pi)^4\delta_4(p + p' - k - k')}{4|p\cdot k|}$$

$$\times \tfrac{1}{4}\sum_{s,t}\left|\mathcal{M}(\{(-p', s'), (-k', t')\}, \{(p, s), (k, t)\})\right|^2,$$

(17.40)

where, as for electron–electron scattering, we will discuss unpolarised cross sections. This requires averaging over the polarisations of the incoming particles (at the end of this chapter we will mention the dependence on the photon polarisations). Note that the photon has also two helicity eigenstates, together with the electron $\sum_{s,t}$ contains four terms. The reduced matrix element for the two diagrams is given by

$$\mathcal{M} = \frac{\bar{u}_{s'}(p')e\gamma^\mu \varepsilon_\mu^{(t')}(k')^*(\slashed{p}+\slashed{k}+m)e\gamma^\nu \varepsilon_\nu^{(t)}(k)u_s(p)}{(p+k)^2 - m^2 + i\varepsilon} + ((k,t) \longleftrightarrow (-k',t')).$$

(17.41)

We leave it as an exercise to verify that the cut photon propagator, for the choice of polarisations discussed in Equation (16.21), is given by ($k^2 \equiv 0$)

$$\equiv \sum_{t=\pm} \varepsilon_\mu^{(t)}(k)\varepsilon_\nu^{(t)}(k)^* = -\left(g_{\mu\nu} - \frac{k_\mu n_\nu(k) + k_\nu n_\mu(k)}{k \cdot n(k)}\right).$$

(17.42)

Like for electron–electron scattering, we can compute $|\mathcal{M}|^2$ graphically by

$$= e^4 \left\{ \frac{\text{Tr}\left[\gamma^\mu(\slashed{p}+\slashed{k}+m)\gamma^\nu(\slashed{p}+m)\gamma^{\nu'}(\slashed{p}+\slashed{k}+m)\gamma^{\mu'}(\slashed{p}'+m)\right]}{\left((p+k)^2 - m^2 + i\varepsilon\right)^2} \right.$$

$$\left. + \frac{\text{Tr}\left[\gamma^\mu(\slashed{p}+\slashed{k}+m)\gamma^\nu(\slashed{p}+m)\gamma^{\mu'}(\slashed{p}-\slashed{k}'+m)\gamma^{\nu'}(\slashed{p}'+m)\right]}{\left((p+k)^2 - m^2 + i\varepsilon\right)\left((p-k')^2 - m^2 + i\varepsilon\right)} \right\}$$

$$\times \sum_{t'=\pm} \varepsilon_\mu^{(t')}(k')\varepsilon_{\mu'}^{(t')}(k')^* \sum_{t=\pm} \varepsilon_{\nu'}^{(t)}(k)\varepsilon_\nu^{(t)}(k)^* + (k' \longleftrightarrow -k).$$

(17.43)

The gauge invariance (i.e., conservation of the fermionic current) is again instrumental in decoupling the longitudinal component of the photon field. In this case the argument is somewhat more subtle. Consider, for example, the term from the cut photon propagators that contains k_ν. It gives rise to the combination (using that $p^2 = m^2$)

$$(\slashed{p}+\slashed{k}+m)\gamma^\nu k_\nu(\slashed{p}+m) = (\slashed{p}+\slashed{k}+m)\{(\slashed{p}+\slashed{k}-m) - (\slashed{p}-m)\}(\slashed{p}+m)$$
$$= ((p+k)^2 - m^2)(\slashed{p}+m).$$

(17.44)

This means that one of the photon vertices is removed. There remain two diagrams, each with one fermion loop and with an odd number of vertices. Furry's theorem tells us that these two diagrams add to zero. We may therefore just as well replace the cut photon propagator by $-g_{\mu\nu}$. Using this we find

$$\sum_{s_1,s_2,t_1,t_2} |\mathcal{M}|^2 = e^4 \left\{ \frac{\mathrm{Tr}\left[\gamma^\mu(\not p + \not k + m)\gamma^\nu(\not p + m)\gamma_\nu(\not p + \not k + m)\gamma_\mu(\not p' + m)\right]}{\left((p+k)^2 - m^2 + i\varepsilon\right)^2} \right.$$

$$\left. + \frac{\mathrm{Tr}\left[\gamma^\mu(\not p + \not k + m)\gamma^\nu(\not p + m)\gamma_\mu(\not p - \not k' + m)\gamma_\nu(\not p' + m)\right]}{\left((p+k)^2 - m^2 + i\varepsilon\right)\left((p-k')^2 - m^2 + i\varepsilon\right)} \right\}$$

$$+ (k' \leftrightarrow -k). \tag{17.45}$$

Taking the incoming electron at rest $[p = (m, \vec{0})]$, following similar steps as for electron–electron scattering, one will arrive at the result

$$\frac{d\sigma}{d\Omega} = \frac{\alpha_e^2}{2m^2}\left(\frac{k_0'}{k_0}\right)^2\left(\frac{k_0}{k_0'} + \frac{k_0'}{k_0} - \sin^2\theta\right), \tag{17.46}$$

where θ is the angle of the scattered photon with the direction of the incident photon. From energy and momentum conservation, one finds that

$$k_0' = \frac{k_0}{1 + (k_0/m)(1 - \cos\theta)}. \tag{17.47}$$

For a detailed derivation we refer to Section 5-2-1 of Itzykson and Zuber and to Section 86 of Berestetskii et al. (see Chapter 1 for the reference).

In Itzykson and Zuber, as for most other textbooks, the result is derived by choosing the photon polarisation such that $\varepsilon(k) \cdot p = 0$ [keeping $\varepsilon(k) \cdot k = 0$]. With this choice it is even possible to determine the polarised cross section (the polarisation of the electron is assumed not to be observed)

$$\left(\frac{d\sigma}{d\Omega}\right)_{\mathrm{pol}} = \frac{\alpha_e^2}{4m^2}\left(\frac{k_0'}{k_0}\right)^2\left(\frac{k_0}{k_0'} + \frac{k_0'}{k_0} + 4(\varepsilon' \cdot \varepsilon)^2 - 2\right), \tag{17.48}$$

where ε and ε' are the polarisations of resp. the incident and scattered photon. When $k_0' \ll m$, one obtains the well-known Thomson formula

$$\left(\frac{d\sigma}{d\Omega}\right)_{\mathrm{pol}} = \frac{\alpha_e^2}{m^2}(\varepsilon' \cdot \varepsilon)^2. \tag{17.49}$$

The unpolarised cross section in this limit is obtained by summing over the scattered and averaging over the incident polarisations

$$\frac{d\sigma}{d\Omega} = \frac{\alpha_e^2}{2m^2}(1 + \cos^2\theta) \quad \text{and} \quad \sigma = \frac{8\pi\alpha_e^2}{3m^2}. \tag{17.50}$$

18

Non-Abelian Gauge Theories

DOI: 10.1201/b15364-18

Quantum electrodynamics is an example of a U(1) gauge theory. U(1) is the group of the unimodular complex numbers and determines the transformation of the charged fields

$$\Psi(x) \rightarrow \exp\left(-iq\Lambda(x)\right)\Psi(x) \equiv g(x)\Psi(x). \tag{18.1}$$

It forms a group, which means that for any two elements g, $h \in U(1)$, the product is also in U(1). Furthermore, any element has an inverse g^{-1}, which satisfies $gg^{-1} = g^{-1}g = 1$. The unit 1 satisfies $g1 = 1g = g$, for any $g \in U(1)$. U(1) is called an Abelian group because its product is commutative. For every g, $h \in U(1)$, $gh = hg$.

It is now tempting to generalise this to other, in general, noncommutative groups, which are called non-Abelian groups. It was the way how Yang and Mills discovered SU(2) gauge theories in 1954. Like for U(1) gauge theories, they made the SU(2) transformation into a local one, where at every point the field can be transformed independently. (It should be noted that they were originally after describing the isospin symmetry that relates protons to neutrons, which form a so-called isospin doublet.)

The simplest non-Abelian gauge group, for which no longer $gh = hg$, is SU(2). This group is well known from the description of spin one-half particles. It has a two-dimensional (spinor) representation, which can also be seen as a representation of the rotation group SO(3). As a local gauge theory, it does no longer act on the spinor indices but on indices related to some internal space, giving rise to so-called internal symmetries. The way the gauge group G acts on the fields Ψ is described by a representation of the group G. A representation defines a mapping ρ from G to the space of linear mappings Map(V), of the linear vector space V into itself:

$$\rho : G \rightarrow \text{Map}(V), \quad \rho(g) : V \rightarrow V. \tag{18.2}$$

Mostly, V will be either $\mathbb{R}^n$ or $\mathbb{C}^n$, in which case $\rho(g)$ is resp. a real or a complex $n \times n$ matrix. For ρ to be a representation, it has to preserve the group structure of G

$$\rho(g)\rho(h) = \rho(gh), \quad \rho(1) = \text{id}_V. \tag{18.3}$$

We will generally restrict the gauge symmetries to Lie groups for which one can write any group element as an exponential of a Lie algebra element

$$g \equiv \exp(X), \quad X \in L_G. \tag{18.4}$$

This Lie algebra has a noncommutative, antisymmetric bilinear product [required to satisfy the Jabobi identity, as defined in eq. (18.12)]

$$(X, Y) \in L_G \times L_G \rightarrow [X, Y] \in L_G. \tag{18.5}$$

The Campbell–Baker–Hausdorff formula expresses that the logarithm of $\exp(X)\exp(Y)$ is an element of the Lie algebra, i.e., the product of two exponentials is again an exponential.

$$F(X, Y) \equiv \log\left(\exp(X)\exp(Y)\right) = X + Y + \tfrac{1}{2}[X, Y] + \tfrac{1}{12}\big[X, [X, Y]\big]$$
$$+ \tfrac{1}{12}\big[Y, [Y, X]\big] + \cdots \in L_G. \tag{18.6}$$

This formula will be of great help in finding a simple criterion for ρ to be a representation, satisfying eq. (18.3). Apart from the group structure of $\mathrm{Map}(V)$, it also has a Lie algebra structure (the commutator of two $n \times n$ matrices is again an $n \times n$ matrix). The representations of the group can be easily restricted to the Lie algebra

$$\rho: L_G \rightarrow \mathrm{Map}(V), \tag{18.7}$$

in a way that preserves the Lie algebra structure

$$\rho([X, Y]) = [\rho(X), \rho(Y)] = \rho(X)\rho(Y) - \rho(Y)\rho(X). \tag{18.8}$$

It is more or less by construction that we require

$$\rho\left(\exp(X)\right) = \exp\left(\rho(X)\right), \tag{18.9}$$

where on the left-hand side ρ is the group representation and on the right-hand side it is the Lie algebra representation. Without causing too much confusion, we can use the same symbol for the two objects. As a Lie algebra forms a linear vector space, we can define a basis on L_G

$$Z = \sum_{a=1}^{n} z_a T^a \in L_G, \quad z_a \in \mathbb{R} \,(\text{or}\,\mathbb{C}), \quad T^a \in L_G. \tag{18.10}$$

In here n is the dimension of the Lie algebra (and the Lie group if, as we will assume throughout, the exponential is locally an invertible mapping). The commutator, also called a Lie product, is completely determined by the structure constants f_{abc}

$$[T^a, T^b] = \sum_{c} f_{abc} T^c. \tag{18.11}$$

Using the Jacobi identity

$$[X, [Y, Z]] + [Y, [Z, X]] + [Z, [X, Y]] = 0, \qquad (18.12)$$

applied to $X = T^a$, $Y = T^b$ and $Z = T^c$, we find (from now on sums over repeated group indices are implicit)

$$f_{bcd}f_{ade} + f_{cad}f_{bde} + f_{abd}f_{cde} = 0. \qquad (18.13)$$

This precisely coincides with the commutation relations of the so-called adjoint representation

$$\left(T^a_{\mathrm{ad}}\right)_{bc} \equiv \rho_{\mathrm{ad}}(T^a)_{bc} = f_{acb}. \qquad (18.14)$$

Indeed, one easily verifies that

$$[\rho_{\mathrm{ad}}(T^a), \rho_{\mathrm{ad}}(T^b)] = f_{abc}\rho_{\mathrm{ad}}(T^c). \qquad (18.15)$$

In general, since a representation preserves the commutation relations, it also preserves the structure constants in terms of $\rho(T^a) \equiv T^a_\rho$, which forms a basis for the linear representation space which is contained in V. They are called the generators of the representation. With the help of eq. (18.6), we easily verify that ρ is a representation if and only if

$$[T^a_\rho, T^b_\rho] = f_{abc} T^c_\rho. \qquad (18.16)$$

This is because under the action of ρ one simply replaces T^a by T^a_ρ

$$\rho\left(\exp(x_a T^a)\right) = \rho\left(\exp(X)\right) = \exp\left(\rho(X)\right) = \exp\left(x_a T^a_\rho\right). \qquad (18.17)$$

Similarly, the Campbell–Baker–Hausdorff formula, when expressed with respect to the Lie algebra basis $\{T^a\}$

$$\exp\left(x_a T^a\right)\exp\left(y_b T^b\right) = \exp\left(\{x_a + y_a + \tfrac{1}{2}x_b y_c f_{bca} + \tfrac{1}{12}(x_d x_b y_c\right.$$
$$\left. + y_d y_b x_c) f_{bce} f_{dea} + \cdots\}T^a\right). \qquad (18.18)$$

directly determines the multiplication of the representation of group elements by replacing T^a by T^a_ρ, provided eq. (18.16) is satisfied. Note that the structure constants are antisymmetric with respect to the first two indices. They are also invariant under cyclic permutations of the indices. This follows from the cyclic property of the trace

$$f_{abd}\mathrm{Tr}(T^d_\rho T^c_\rho) = \mathrm{Tr}([T^a_\rho, T^b_\rho]T^c_\rho) = \mathrm{Tr}(T^a_\rho[T^b_\rho, T^c_\rho]) = f_{bcd}\mathrm{Tr}(T^a_\rho T^d_\rho), \qquad (18.19)$$

and from the fact that for compact groups the generators can be normalised such that

$$\mathrm{Tr}(T^a_{\mathrm{fnd}} T^b_{\mathrm{fnd}}) = -\tfrac{1}{2}\delta_{ab}, \qquad (18.20)$$

where T^a_{fnd} are the generators of the so-called fundamental or defining representation of the group G. This matrix representation is usually identified with the group (or algebra) itself, which till now was seen more as an abstract entity. The simplest example is SU(2), the set of complex unitary 2×2 matrices with unit determinant. Its fundamental representation coincides with the spinor or spin one-half representation. The structure constants and the generators of the fundamental and adjoint representations were considered in Chapter 12 [see eq. (12.9)]

$$T^a_{\text{fnd}} = -\frac{i}{2}\sigma^a, \quad f_{abc} = \varepsilon_{abc}, \quad \rho_{ad}(T^a) = -L^a. \tag{18.21}$$

Because the Campbell–Baker–Hausdorff formula plays such a crucial role in the theory and in the practical implementation of group representations, we will now provide a more abstract derivation of eq. (18.6) to all orders. The proof simply states how the Taylor expansion products of Lie algebra elements are regrouped in multiple commutators. A crucial ingredient for deriving the Campbell–Baker–Hausdorff is the so-called *derivation* $\mathcal{D}$, that maps a product of Lie algebra elements into a multiple commutator.

$$\mathcal{D}X = X, \quad \mathcal{D}X_{i_1} X_{i_2} \cdots X_{i_s} \equiv \left[X_{i_1}, [X_{i_2}, \cdots [X_{i_{s-1}}, X_{i_s}]\cdots]\right], \quad s > 1. \tag{18.22}$$

We also define for these products the adjoint map, ad, introduced in eq. (12.12)

$$\text{ad}\, X_{i_1} X_{i_2} \cdots X_{i_s} \equiv \text{ad}\, X_{i_1} \text{ad}\, X_{i_2} \cdots \text{ad}\, X_{i_s}, \tag{18.23}$$

which is easily seen to satisfy

$$\text{ad}([X, Y]) = [\text{ad}\, X, \text{ad}\, Y]. \tag{18.24}$$

It is more or less by definition that for any two products u and v of Lie algebra elements

$$\mathcal{D}(uv) = \text{ad}\, u \mathcal{D} v. \tag{18.25}$$

For two Lie algebra elements X and Y, it can easily be shown that

$$\mathcal{D}[X, Y] = \mathcal{D}(XY) - \mathcal{D}(YX) = \text{ad}\, X \mathcal{D} Y - \text{ad}\, Y \mathcal{D} X$$
$$= [X, \mathcal{D} Y] + [\mathcal{D} X, Y] \tag{18.26}$$

and this allows us to prove by induction that a monomial Q (a polynomial of which all terms are of the same order) of degree m in terms of Lie algebra elements $X_i, i = 1, 2, \ldots, s$ is an element of the Lie algebra (i.e., can be written as multiple commutators, called a Lie monomial) if and only if $\mathcal{D}Q = mQ$. If this equation is satisfied, it is clear from the definition of a derivation that Q is a Lie monomial. So it is sufficient to prove that the equation is satisfied for Q as a Lie monomial. In that case Q is a sum of terms, each of which can be written as $\text{ad}(X_{i_1})Q^{(1)}$ with $Q^{(1)}$ a Lie monomial of degree $m - 1$. Using eq. (18.26) therefore yields $\mathcal{D}\mathrm{ad}(X_{i_1})Q^{(1)} = \mathrm{ad}(X_{i_1})\mathcal{D}Q^{(1)} + \mathrm{ad}(X_{i_1})Q^{(1)}$. Induction in m gives the required result.

Now it is trivial to regroup the terms in the Taylor expansion of eq. (18.6) in multiple commutators. From the fact that any group element can be written as the exponent of a Lie algebra element, we know that $F(X, Y) \in L_G$ (at the worst one needs to restrict X and Y to sufficiently small neighbourhoods of the origin in L_G). Consequently, in the Taylor expansion of $F(X, Y)$, the collection of all terms at fixed order m, denoted by $F_m(X, Y)$, is a monomial in X and Y, and $F_m(X, Y)$ is an element of the Lie algebra such that

$$F(X, Y) \equiv \sum_m F_m(X, Y), \quad F_m(X, Y) = \frac{1}{m}\mathcal{D}F_m(X, Y). \tag{18.27}$$

It is not difficult to work out the Taylor expansion for $F(X, Y)$

$$F(X, Y) \equiv \log\left(\exp(X)\exp(Y)\right) = \log\left(1 + \sum_{i+j>0} \frac{X^i Y^j}{i!\,j!}\right)$$

$$= \sum_k \frac{(-1)^{k-1}}{k}\left(\sum_{i+j>0} \frac{X^i Y^j}{i!\,j!}\right)^k, \tag{18.28}$$

from which we easily obtain the explicit expression for the Campbell–Baker–Hausdorff formula in terms of multiple commutators,

$$F(X, Y) = \sum_m \sum_{\{k,\,\sum_{j=1}^k p_j+q_j=m,\; p_j+q_j>0\}} \frac{(-1)^{k-1}}{km}\frac{\mathcal{D}\left(X^{p_1}Y^{q_1}X^{p_2}Y^{q_2}\cdots X^{p_k}Y^{q_k}\right)}{p_1!q_1!p_2!q_2!\cdots p_k!q_k!}.$$

$$\tag{18.29}$$

We leave it to the industrious student to verify that

$F_1(X, Y) = X + Y, \quad F_2(X, Y) = \frac{1}{2}[X, Y],$

$F_3(X, Y) = \frac{1}{12}\left\{(\mathrm{ad}X)^2 Y + (\mathrm{ad}Y)^2 X\right\}, \; F_4(X, Y) = -\frac{1}{24}\mathrm{ad}X\mathrm{ad}Y\mathrm{ad}X(Y),$

$F_5(X, Y) = -\frac{1}{720}\left\{(\mathrm{ad}X)^4 Y + (\mathrm{ad}Y)^4 X\right\} + \frac{1}{360}\left\{\mathrm{ad}X(\mathrm{ad}Y)^3 X + \mathrm{ad}Y(\mathrm{ad}X)^3 Y\right\}$

$$\qquad - \frac{1}{120}\left\{\mathrm{ad}X\mathrm{ad}Y(\mathrm{ad}X)^2 Y + \mathrm{ad}Y\mathrm{ad}X(\mathrm{ad}Y)^3 X\right\}. \tag{18.30}$$

After this intermezzo we return to the issue of constructing non-Abelian gauge theories. The simplest way is by generalising first the covariant derivative. U(1) gauge transformations act on a complex field as in eq. (18.1), and the covariant derivative is designed such that

$$D_\mu \Psi(x) \rightarrow g(x)D_\mu \Psi(x). \tag{18.31}$$

Since the gauge field transforms as in eq. (17.3), this is easily seen to imply that the covariant derivative is defined as in eq. (17.2) [these formula are of course also valid for complex scalar fields; compare this to eq. (3.36)]. For a non-Abelian gauge theory, we consider first a field Ψ that transforms as an irreducible representation (i.e., there is no *nontrivial* linear subspace that is left invariant under the action of *all* gauge group elements)

$$\Psi \rightarrow {}^g\Psi \equiv \rho(g)\Psi. \tag{18.32}$$

In the following, as in the literature, we shall no longer make a distinction between g and $\rho(g)$. It will always be clear from the context what is intended. The vector potential should now be an element of the Lie algebra L_G (more precisely a representation thereof)

$$A_\mu = A_\mu^a T^a. \tag{18.33}$$

For U(1), which is one dimensional, we need to define $T^1 \equiv i$. The Lie algebra of the group consisting of the unimodular complex numbers is the set of imaginary numbers $L_{U(1)} = i\mathbb{R}$. Note that as an exception this generator is normalised different from eq. (18.20), so as not to introduce unconventional normalisations elsewhere. The real valued vector potential A_μ will now be denoted by A_μ^1 and we see that under a gauge transformation

$$A_\mu \rightarrow {}^gA_\mu = gA_\mu g^{-1} - q^{-1}(\partial_\mu g)g^{-1} = gA_\mu g^{-1} + q^{-1}g\partial_\mu(g^{-1}). \tag{18.34}$$

This is the form that generalises directly to the non-Abelian gauge groups with the covariant derivative defined by

$$D_\mu\Psi = \left(\partial_\mu + qA_\mu\right)\Psi, \tag{18.35}$$

where $A_\mu \equiv A_\mu^a T_\rho^a$ is a matrix acting on the fields Ψ. We leave it as an exercise to verify that under a gauge transformation, eq. (18.31) remains valid for the non-Abelian case.

It is now a trivial matter to construct a Lagrangian that is invariant under local gauge transformation. Assuming the representation is unitary, for a scalar field Ψ one has

$$\mathcal{L}_\Psi = \left(D_\mu\Psi\right)^\dagger D^\mu\Psi - m^2\Psi^\dagger\Psi, \tag{18.36}$$

whereas if Ψ is a Dirac field, carrying both spinor (representation of the Lorentz group) and group indices, one has

$$\mathcal{L}_\Psi = \overline{\Psi}\left(i\gamma^\mu D_\mu - m\right)\Psi, \quad \overline{\Psi} = \Psi^\dagger\gamma^0, \tag{18.37}$$

where $\Psi^\dagger$ is the Hermitian conjugate both with respect to the spinor and the group (representation) indices.

The part of the Lagrangian that describes the self-interactions of the vector field A_μ has to be invariant under local gauge transformations too. In that respect U(1) or Abelian gauge theories are special, since the homogeneous part of the transformation of the vector potential is trivial, $gA_\mu g^{-1} = A_\mu$. For non-Abelian gauge transformations, this is no longer true. For U(1) one easily verifies that

$$D_\mu D_\nu \Psi - D_\nu D_\mu \Psi \equiv [D_\mu, D_\nu]\Psi = iq F^1_{\mu\nu}\Psi, \qquad (18.38)$$

where $F^1_{\mu\nu} = \partial_\mu A^1_\nu - \partial_\nu A^1_\mu$ is the electromagnetic field strength; compare this to eq. (3.27). Because the covariant derivative transforms in a simple way under gauge transformations, this formula can be directly generalised to non-Abelian gauge theories

$$F_{\mu\nu} \equiv q^{-1}[D_\mu, D_\nu] \overset{g}{\to} g F_{\mu\nu} g^{-1}. \qquad (18.39)$$

For U(1), where g is a number, this means that the field strength is gauge invariant, as was noted before. For non-Abelian gauge theories the field strength itself is not gauge invariant. Nevertheless, it is simple to construct a gauge-invariant action for the gauge field

$$\mathcal{L}_A = \tfrac{1}{2}\mathrm{Tr}\left(F_{\mu\nu}F^{\mu\nu}\right) \equiv -\tfrac{1}{4}F^a_{\mu\nu}F_a^{\mu\nu}, \qquad (18.40)$$

where $F^a_{\mu\nu}$ are the components of the field strength with respect to the Lie algebra basis,

$$F_{\mu\nu} \equiv F^a_{\mu\nu}T^a = \left(\partial_\mu A^a_\nu - \partial_\nu A^a_\mu + q f_{abc} A^b_\mu A^c_\nu\right)T^a. \qquad (18.41)$$

We see from $\mathcal{L}_A$ and $\mathcal{L}_\Psi$ that q plays the role of an expansion parameter. For $q = 0$ we have $n = \dim(G)$ noninteracting photon fields. They couple with strength q to the scalar or Dirac fields. For non-Abelian gauge theories, in addition the vector field couples to itself. These *self-interactions* guarantee that there is invariance under the gauge group G, which is much bigger than $U(1)^n$, which is the symmetry that seems implied by the $q = 0$ limit. The non-Abelian gauge invariance fixes the "charges" of the fields with respect to each of these U(1) gauge factors. Without the non-Abelian gauge symmetry, there would have been n independent 'charges.'

The Lagrangian $\mathcal{L}_A$ is the one that was discovered in 1954 by C.N. Yang and R.L. Mills. The Euler–Lagrange equations for the Lagrangian $\mathcal{L}_A$ are called the Yang–Mills equations. One easily shows that

$$\partial_\mu F_a^{\mu\nu} + q f_{abc} A^b_\mu F_c^{\mu\nu} = 0 \quad \text{or} \quad [D_\mu, F^{\mu\nu}] \equiv \partial_\mu F^{\mu\nu} + q[A_\mu, F^{\mu\nu}] = 0. \quad (18.42)$$

For the coupling to fermions we read off from eq. (18.37) what the current for the Yang–Mills field is

$$\mathcal{L}_\Psi = \overline{\Psi}(i\gamma^\mu D_\mu - m)\Psi = \overline{\Psi}(i\gamma^\mu \partial_\mu - m)\Psi + iq A^a_\mu \overline{\Psi}\gamma^\mu T^a \Psi. \quad (18.43)$$

The current is therefore given by

$$J^a_\mu \equiv -iq\overline{\Psi}\gamma_\mu T^a \Psi. \qquad (18.44)$$

The coupled Yang–Mills equations read

$$\partial_\mu F_a^{\mu\nu} + q f_{abc} A_\mu^b F_c^{\mu\nu} = J_a^\nu. \qquad (18.45)$$

In Problem 31 it will be shown that the current is not gauge invariant, unlike for Abelian gauge symmetries. Closely related is the fact that it is no longer true that the current is conserved, i.e., $\partial^\mu J_\mu^a \neq 0$. Instead, it will be shown in Problem 31 that $\partial^\mu J_\mu^a + q f_{abc} A_b^\mu J_\mu^c = 0$.

19

The Higgs Mechanism

DOI: 10.1201/b15364-19

We have seen in Problem 30 that the four-Fermi interaction in good approximation can be written in terms of the exchange of a heavy vector particle. In lowest order we have resp. the diagrams in Figures 19.1(a) and 19.1(b). The first diagram comes from a four-fermion interaction term that can be written in terms of the product of two currents $J_\mu J^\mu$, where $J_\mu = \overline{\Psi}\gamma_\mu\Psi$. Here each fermion line typically carries its own flavour index, which was suppressed for simplicity. Figure 19.1(b) can be seen to effectively correspond to

$$-\tilde{J}^\mu(-k)\left(\frac{g_{\mu\nu} - k_\mu k_\nu/M^2}{k^2 - M^2 + i\varepsilon}\right)\tilde{J}^\nu(k). \tag{19.1}$$

At values of the exchanged momentum $k^2 \ll M^2$, one will not see a difference between these two processes, provided the coupling constant for the four-Fermi interactions [Figure 19.1(a)] is chosen suitably (see Problem 30). This is because for small k^2, the propagator can be replaced by $g_{\mu\nu}/M^2$, which indeed converts eq. (19.1) to $J^\mu J_\mu/M^2$. It shows that the four-Fermi coupling constant is proportional to M^{-2}, such that its weakness is explained by the heavy mass of the vector particle that mediates the interactions. Examples of four-Fermi interactions occur in the theory of β-decay, for example, the decay of a neutron into a proton, an electron and an antineutrino. In that case the current also contains a γ^5 (Problem 40).

It turns out that the four-Fermi theory cannot be renormalised. Its quantum corrections give rise to an infinite number of divergent terms that cannot be reabsorbed in a redefinition of a Lagrangian with a finite number of interactions. With the interaction resolved at higher energies by the exchange of a massive vector particle, the situation is considerably better. But it becomes crucial for the currents in question to be conserved, such that the $k_\mu k_\nu$ part in the propagator has no effect. It would give rise to violations of unitarity in the scattering matrix at high energies [the σ field defined in eq. (16.15) has the wrong sign for its kinetic part]. To enforce current conservation, we typically use gauge invariance. But gauge invariance would protect the vector particle from having a mass. The big puzzle therefore was how to describe a massive vector particle that is nevertheless associated to the vector potential of a gauge field.

(a)   (b)

FIGURE 19.1
Lowest order diagrams.

The answer can be found in the theory of superconductivity, which prevents magnetic field lines from penetrating in a superconducting sample. If there is, however, penetration in the form of a quantised flux tube, the magnetic field decays exponentially outside the flux tube. This would indicate a mass term for the electromagnetic field within the superconductor. The Landau–Ginzburg theory that gives an effective description of this phenomenon [the microscopic description being given by the Bardeen–Cooper–Schrieffer (BCS) theory of Cooper pairs] precisely coincides with scalar quantum electrodynamics.

$$\mathcal{L} = -\tfrac{1}{4}F_{\mu\nu}F^{\mu\nu} + \left(D_\mu\varphi\right)^* D^\mu\varphi - \kappa\varphi^*\varphi - \tfrac{1}{4}\lambda\left(\varphi^*\varphi\right)^2. \qquad (19.2)$$

In the Landau–Ginzburg theory, φ describes the Cooper pairs. It is also called the order parameter of the BCS theory. In usual scalar quantum electrodynamics, we would put $\kappa = m^2$, where m is the mass of the charged scalar field. But in the Landau–Ginzburg theory of superconductivity, it happens to be the case that κ is *negative*. In that case the potential $V(\varphi)$ for the scalar field has the shape of a Mexican hat, Figure 19.2.

The minimum of the potential is no longer at $\varphi = 0$, but at $\varphi^*\varphi = -2\kappa/\lambda$, and is *independent* of the phase of φ. To find the physical excitations of this theory, we have to expand around the minimum. With a global phase rotation we can choose the point to expand around to be real,

$$\varphi_0 = \sqrt{-2\kappa/\lambda}. \qquad (19.3)$$

But this immediately implies that the quadratic terms in the gauge field give rise to a mass term for the photon field

$$|D_\mu\varphi_0|^2 = e^2\varphi_0^2 A_\mu A^\mu \equiv \tfrac{1}{2}M^2 A_\mu A^\mu, \quad M = 2e\sqrt{-\kappa/\lambda}. \qquad (19.4)$$

Furthermore, from the degeneracy of the minimum of the potential it follows that the fluctuation along that minimum [the phase in $\varphi = \varphi_0 \exp(i\chi)$] has no mass (this is related to the famous Goldstone theorem, which states that if choosing a minimum of the potential would break the symmetry, called *spontaneous* symmetry breaking, there is always a massless particle). However, this phase χ is precisely related to the gauge invariance and can be rotated away by a gauge transformation. On the one hand χ corresponds to a massless excitation; on the other hand it is the unphysical longitudinal component of the gauge field. But the photon becomes massive and has to develop an

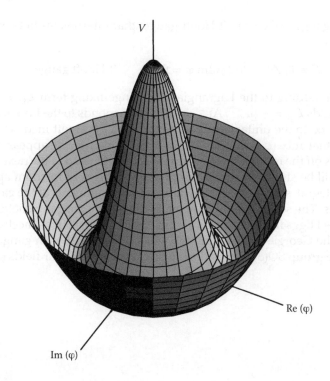

FIGURE 19.2
The Mexican hat potential $V(\varphi)$.

additional physical polarisation, which is precisely the longitudinal component. In a prosaic way one states that the massless excitation (called a *would-be*
Goldstone boson) was 'eaten' by the longitudinal component of the photon,
which in the process got a mass ('got fat').

This means we have four massive degrees of freedom, three for the massive
vector particle and one for the absolute value of the complex scalar field
(its mass is determined by the quadratic part of the potential in the radial
direction at $\varphi = \varphi_0$). This is exactly the same number as for ordinary scalar
electrodynamics where $\kappa > 0$, because in that case the massless photon has
only two degrees of freedom, whereas the complex scalar field represents two
massive real scalar fields. It looks, however, like there is a discontinuity in the
description of these degrees of freedom when approaching $\kappa = 0$. But the
interpretation of the phase of the complex field as a longitudinal component
of the vector field is simply a matter of choosing a particular gauge. To count
the number of degrees of freedom, we implicitly made two different gauge
choices

$$\kappa > 0: \quad \partial_\mu A^\mu = 0, \quad \text{Lorentz gauge,}$$

$$\kappa < 0: \quad \text{Im } \varphi = 0, \quad \text{Unitary gauge.} \tag{19.5}$$

There is a gauge, called the 't Hooft gauge, that interpolates between these two gauges

$$\mathcal{F} \equiv \partial_\mu A^\mu - 2ie\xi\varphi_0 \mathrm{Im}\,\varphi = 0, \qquad \text{'t Hooft gauge.} \qquad (19.6)$$

Rather than adding to the Lagrangian the gauge-fixing term $\mathcal{L}_{\text{gf}} = -\frac{1}{2}\alpha(\partial_\mu A^\mu)^2$, one adds $\mathcal{L}_{\text{gf}} = -\frac{1}{2}\alpha\mathcal{F}^2$. At $\xi = 0$ this corresponds to the Lorentz gauge, and at $\xi = \infty$ to the unitary gauge. For the choice 't Hooft made ($\xi = 1/\alpha$), the terms that mix ($\varphi - \varphi_0$) and A_μ at quadratic order disappear and one easily reads off the masses. Gauge fixing will be discussed in the next chapter, where it will be shown how extra unphysical degrees of freedom appear in the path integral, so as to cancel the unphysical components of the gauge and scalar fields. The scalar field, whose interactions give the gauge field a mass, is called the Higgs field. Problems 34 and 35 discuss the Higgs mechanism in detail for the Georgi–Glashow model, which is a non-Abelian gauge theory with gauge group SO(3), coupled to an SO(3) vector of scalar fields φ^a.

20

Gauge Fixing and Ghosts

DOI: 10.1201/b15364-20

The quantisation of gauge theories in the path integral formalism requires more discussion, since the gauge condition (like the Lorentz gauge $\partial_\mu A^\mu = 0$) seems to remove only one degree of freedom of the two that are eliminated in the Hamiltonian formulation (see Chapter 16). From a simple example it is easily demonstrated what the effect of gauge fixing on a (path) integral is. For this we take $f(\vec{x})$ to be a function on $\mathbb{R}^3$, which is invariant under rotations around the origin, such that it is a function $f(r)$ of the radius $r = |\vec{x}|$ only. The symmetry group is hence $SO(3)$, and we can attempt to compute the integral $\int d_3\vec{x}\, f(\vec{x})$ by introducing a 'gauge' fixing condition like $x_2 = x_3 = 0$. But it is clear that

$$\int d_3\vec{x}\, f(\vec{x}) \neq \int d_3\vec{x}\, \delta(x_2)\delta(x_3) f(\vec{x}) = \int dx_1\, f(x_1). \tag{20.1}$$

We know very well that we need a Jacobian factor for the radial integral

$$\int d_3\vec{x}\, f(\vec{x}) = 4\pi \int_0^\infty r^2 dr\, f(r). \tag{20.2}$$

This Jacobian, arising in the change of variable to the invariant radial coordinates and the angular coordinates, can be properly incorporated following the method introduced by Faddeev and Popov. The starting point is a straightforward generalisation of the identity $\int dx\, |f'(x)|\delta[f(x)] = 1$, assuming the equation $f(x) = 0$ to have precisely one solution (in a sense the right-hand side of the equation counts the number of zeros). It reads

$$1 = \int \mathcal{D}g \,|\det\left(M(^g A)\right)|\delta\left(\mathcal{F}(^g A)\right), \tag{20.3}$$

where $\mathcal{F}(A) \in L_G$ is the gauge-fixing function [with the gauge condition $\mathcal{F}(A) = 0$, e.g., $\mathcal{F}(A) = \partial_\mu A^\mu = 0$]. The gauge transformation g of the gauge field A is indicated by $^g A$; see Equation (18.34). Furthermore, $M(A) : L_G \to L_G$ plays the role of the Jacobian,

$$M(^g A) = \frac{\partial \mathcal{F}(^g A)}{\partial g} \equiv \frac{\partial \mathcal{F}(e^{X}{}_g A)}{\partial X} \quad \text{at } X = 0. \tag{20.4}$$

Equivalently, with respect to the Lie algebra basis, where $\mathcal{F}(A) \equiv \mathcal{F}_a(A)T^a$, one has

$$M_{ab}(A) = \frac{d\mathcal{F}_a(^{\exp(tT^b)}A)}{dt} \quad \text{at} \quad t = 0. \tag{20.5}$$

To relate this to the previous equation, one makes use of the fact that

$$^h(^g A) = {}^{(hg)}A, \tag{20.6}$$

which states that two successive gauge transformations, g and h, give the same result as a single gauge transformation with hg.

As an example we consider the Lorentz gauge, with $\mathcal{F}(A) = \partial_\mu A^\mu$, for which

$$\mathcal{F}(^{\exp(X)}A) - \mathcal{F}(A) = -q^{-1}\partial_\mu D^\mu_{\text{ad}}(A)(X) + \mathcal{O}(X^2), \tag{20.7}$$

where $D^\mu_{\text{ad}}(A)$ is the covariant derivative in the adjoint representation

$$D^\mu_{\text{ad}}(A)(X) \equiv \partial^\mu X + q[A^\mu, X]. \tag{20.8}$$

With respect to the Lie algebra basis, this gives

$$q\,M_{ab}(A) = -\delta_{ab}\partial_\mu\partial^\mu + qf_{abc}(\partial^\mu A^c_\mu + A^c_\mu\partial^\mu). \tag{20.9}$$

For an Abelian gauge theory, the structure constants f_{abc} vanish and $M(A)$ becomes independent of the gauge field. This means that $\det[M(A)]$ can be absorbed in an overall normalisation of the path integral. For non-Abelian gauge theories this is no longer possible. Before describing how the A dependence of $\det[M(A)]$ is incorporated, it is important to note that we assumed the gauge condition $\mathcal{F}(^g A) = 0$ to have precisely one solution, which can be arranged with the help of Equation (20.6) to occur at $g = 1$, in which case A is said to satisfy the gauge condition. This is, in general, not correct, as was discovered by Gribov. Even in our simple problem on $\mathbb{R}^3$, the gauge condition $x_2 = x_3 = 0$ does not uniquely specify the gauge, because we can go from $(r, 0, 0)$ to $(-r, 0, 0)$ through a rotation over 180 degrees. We have to introduce a further restriction to get the identity

$$\int d_3\vec{x}\, f(\vec{x}) = 4\pi \int d_3\vec{x}\, x_1^2\, \delta(x_2)\delta(x_3)\theta(x_1)f(\vec{x}), \tag{20.10}$$

where $\theta(x) = 0$ for $x < 0$ and $\theta(x) = 1$ for $x \geq 0$. In perturbation theory only the gauge fields near the origin in field space are relevant, and gauge conditions are chosen so as to avoid this problem in a small neighbourhood of the origin. The Lorentz gauge is such a gauge condition, and the gauge fixing or Gribov ambiguity is not an issue for computing quantities in perturbation theory in q.

We still need to define what we mean by $\mathcal{D}g$ in Equation (20.3). It stands for the integration measure $\prod_x dg(x)$, with $dg(x)$ for every x defined as the so-called Haar measure on the group. It is best described in the example of SU(2), which as a space is isomorphic with S^3. When S^3 is embedded in $\mathbb{R}^4$ as a unit sphere, $n_\mu^2 = 1$, it is not too difficult to see that $g = n_4 + i\sigma_k n_k$ gives an element of SU(2), whereas $\exp(i\chi s_k \sigma_k) = \cos(\chi) + i\sin(\chi)s_k\sigma_k$, with $s_k^2 = 1$ shows that any element of SU(2) can be written in terms of n_μ. The Haar measure coincides with the standard integration measure on S^3, $\int d_4 n \, \delta(n_\mu^2 - 1)$. The Haar measure is, in general, invariant under the change of variables $g \to hg$ and $g \to gh$, for h some fixed group element. We can insert Equation (20.3) in the path integral to obtain

$$Z = \int \mathcal{D}A_\mu \mathcal{D}g \, \det\left(M({}^g A)\right)\delta\left(\mathcal{F}({}^g A)\right) \exp\left(i\,S(A)\right). \tag{20.11}$$

We now use that the action $S(A)$ is invariant under gauge transformations. We leave it as an exercise to verify that likewise $\mathcal{D}A_\mu$ is invariant under the change of variables $A \to {}^g A$, which trivially implies that

$$Z = \int \mathcal{D}A_\mu \mathcal{D}g \, \det\left(M(A)\right)\delta\left(\mathcal{F}(A)\right) \exp\left(i\,S(A)\right). \tag{20.12}$$

The dependence of the integrand on g has disappeared, and the integration over g gives an overall (infinite) normalisation factor, which is irrelevant. We next note that Z has to be independent of the gauge-fixing function $\mathcal{F}$, in particular $\mathcal{F}(A) - Y$ is just as good for the gauge fixing [provided, of course, we show that $\mathcal{F}({}^g A) = Y$ has a solution]. This modification does not affect the so-called Faddeev–Popov operator $M(A)$ and we find

$$Z = \int \mathcal{D}A_\mu \det\left(M(A)\right)\delta\left(\mathcal{F}(A) - Y\right) \exp\left(i\,S(A)\right), \tag{20.13}$$

independent of Y. Suitably normalising $\mathcal{D}Y$ we can define

$$\int \mathcal{D}Y \exp\left(-\frac{\alpha i}{2}\int d_4 x \, Y_a^2(x)\right) = 1, \tag{20.14}$$

which combined with the previous equation gives

$$Z = \int \mathcal{D}Y\mathcal{D}A_\mu \det\left(M(A)\right)\delta\left(\mathcal{F}(A) - Y\right) \exp\left(i\int d_4 x \, \mathcal{L}(A) - \frac{\alpha}{2}Y_a^2(x)\right)$$

$$= \int \mathcal{D}A_\mu \det\left(M(A)\right) \exp\left(i\int d_4 x \, \mathcal{L}(A) - \frac{\alpha}{2}\mathcal{F}_a^2(A)\right). \tag{20.15}$$

For U(1) gauge theories with $\mathcal{F}(A) = \partial_\mu A^\mu$, this precisely reproduces the action of Equation (4.22) in the Lorentz gauge, and in that case $\det[M(A)]$ is a constant.

For non-Abelian gauge theories we are left with the task of computing $\det[M(A)]$ for each A, which is no longer constant. But here the path integral over Grassmann variables comes to the rescue. In Problem 25 we have seen that

$$\int \mathcal{D}\bar{\eta}\mathcal{D}\eta \, \exp\left(i\int d_4x \, \bar{\eta}^a(x)q\, M_{ab}(A)\eta^b(x)\right) = \det\left(M(A)\right), \quad (20.16)$$

up to an overall normalisation. This implies that the path integral can also be written in the Lorentz gauge as

$$Z = \int \mathcal{D}A_\mu \mathcal{D}\bar{\eta}\mathcal{D}\eta \, \exp\left\{i\int d_4x \, \mathcal{L}(A) - \frac{\alpha}{2}\left(\partial_\mu A_a^\mu(x)\right)^2 + \partial^\mu \bar{\eta}^a(x)[\partial_\mu \eta^a(x)\right.$$
$$\left. + q f_{abc} A_\mu^b(x)\eta^c(x)]\right\}. \quad (20.17)$$

Since $\bar{\eta}$ and η are auxiliary fields, they should never appear as external lines. They are therefore called ghosts. Ghosts can only appear in loops and because of the fermionic nature of the ghost variables, every such loop gives a minus sign. The Feynman rules for the Lorentz gauge are given in Table 20.1.

Because one can easily derive that for a complex scalar field (up to an overall constant)

$$\int \mathcal{D}\varphi^* \mathcal{D}\varphi \, \exp\left(i\int d_4x \, \varphi_a^*(x)q\, M^{ab}(A)\varphi_b(x)\right) = \frac{1}{\det\left(M(A)\right)}, \quad (20.18)$$

we can view a ghost as the elimination of a complex degree of freedom. It is in this way that in the path integral the two unphysical degrees of freedom of a Lorentz vector are eliminated. For QED both the ghost and the unphysical degrees of freedom have no interactions and cannot appear as external lines either, which is why in QED the introduction of ghosts was never necessary for a consistent description of the theory. For non-Abelian gauge theories, because of the interaction of the ghost with the gauge field, ghosts can no longer be ignored. To have the ghosts eliminate the unphysical degrees of freedom, one should have the 'masses' (poles) of the ghosts coincide with the 'masses' of the unphysical degrees of freedom. Furthermore the couplings of

TABLE 20.1

Feynman rules for ghosts.

		no external ghost lines
	$(q M)^{-1}(A{=}0)_{ab} = \dfrac{\delta_{ab}}{k^2 + i\varepsilon}$	ghost propagator (Lorentz gauge)
	$-iq f_{abc} k_1^\mu$	ghost vertex (Lorentz gauge)
	$-1 \times i \int \dfrac{d_4k}{(2\pi)^4}$	loop factor

the ghost and unphysical fields to the physical fields should be related. This is verified explicitly for the Georgi–Glashow model in Problem 35. In general it is guaranteed by the existence of an extra symmetry, discovered by Becchi, Rouet, and Stora, called the BRS symmetry s, which, for example, acts on the gauge field as follows:

$$s A^{\mu} = D^{\mu}_{\text{ad}} \eta. \tag{20.19}$$

This is precisely an infinitesimal gauge transformation. For more details, see Itzykson and Zuber, Section 12-4-1.

21

The Standard Model

DOI: 10.1201/b15364-21

The standard model describes the electromagnetic and weak interactions, unified in the so-called electro-weak theory of Glashow–Weinberg–Salam with the gauge group U(1)× SU(2) and the strong interactions, known as quantum chromodynamics (QCD) with gauge group SU(3). Theory and experiment, where tested, agree very well up to about 100 GeV, the energies reached by present-day accelerators. Now that the top quark has been found, at a mass of 174 GeV, only the Higgs particle remains to be detected. Its mass should be smaller than 1000 GeV (i.e., 1 TeV = Terra electronvolt) according to present-day theoretical insight. Gravitation has been left out so far. Its natural scale in energy where quantum effects would become important is the Planck energy, $E_{pl} = \sqrt{\hbar c^5/G} \approx 10^{19}$ GeV. It is very well possible that a number of the fundamental parameters in the standard model will be determined, either directly or indirectly, by gravitational interactions. The standard model should then be considered as an effective field theory. The theory for which the standard model describes its effective low-energy behaviour is called a unified theory. An intermediate stage, which does not yet include gravity is the so-called grand unified theory (GUT). The simplest version unifies the electro-weak and strong interactions using a gauge group SU(5) [which has U(1)×SU(2)×SU(3) as a subgroup], thereby reducing the number of free parameters considerably. These GUTs predict proton decay, albeit at the tremendously low rate of one decay in every 10^{30-31} years. Nevertheless, a swimming pool of $(10\,\text{m})^3$ contains enough protons to verify that the proton decay is slower than can be comfortably accommodated by GUTs. Candidates that unify the standard model with gravity in the form of string theories and supergravity have been unable to provide predictions that either rule them out experimentally or provide evidence in favour of these theories. Much is therefore still to be discovered, in particular because theoretical insight of the last ten years has shown that a Higgs field is most likely not fundamental, although it is not yet ruled out that it will show its structure only at Planck energies. If that is the case, the mass of the Higgs should, however, not be much bigger than 100 GeV.

The standard model consists of gauge fields B_μ [for U(1)], W_μ^a [for SU(2)] and A_μ^a [for SU(3), where a runs from 1 to 8, to be discussed later]. These gauge fields have interactions with a Higgs field $\phi \in \mathbb{C}^2$, which transforms under SU(2) as a spin one-half representation (i.e., the fundamental representation) with a coupling constant g. Under U(1) this Higgs field transforms with a coupling constant $-\frac{1}{2}g'$, whereas it is neutral under SU(3). These couplings are represented in the covariant derivative

$$D_\mu \phi = \partial_\mu \phi - \frac{i}{2} g' B_\mu \phi - \frac{ig}{2} \sum_{a=1}^{3} W_\mu^a \sigma_a \phi. \qquad (21.1)$$

The potential for the Higgs field causes spontaneous breaking of part of the symmetries

$$V(\phi) = \frac{\lambda}{4} (\phi^\dagger \phi - F^2)^2 = \kappa \phi^\dagger \phi + \frac{\lambda}{4} (\phi^\dagger \phi)^2 + \text{const.}, \qquad (21.2)$$

where $\kappa \equiv -\frac{1}{2}\lambda F^2$. In this case the minimum of the potential, also called the vacuum, is degenerate on a three-dimensional sphere, specified by $\phi^\dagger \phi = F^2$ ($\phi \in \mathbb{C}^2 \sim \mathbb{R}^4$), which would give rise to three massless scalar particles according to the Goldstone theorem, but all three will be 'eaten' by longitudinal components of the gauge-fields to which the Higgs field couples. We note that there are four gauge-field components, B_μ and W_μ^a for $a = 1, 2$ and 3. Indeed one combination among these four will not have something to 'eat' and will therefore stay massless. It plays the role of the photon field as we got to know it in QED. To see this, write

$$\phi = \begin{pmatrix} 0 \\ F \end{pmatrix} + \begin{pmatrix} \varphi_1 \\ \varphi_2 \end{pmatrix} \equiv \phi_0 + \begin{pmatrix} \varphi_1 \\ \varphi_2 \end{pmatrix}, \qquad (21.3)$$

such that

$$(D_\mu \phi_0)^\dagger D^\mu \phi_0 = \frac{g^2 F^2}{4} \left[(W_\mu^1)^2 + (W_\mu^2)^2 \right] + \frac{F^2}{4} \left[g' B_\mu - g W_\mu^3 \right]^2. \qquad (21.4)$$

Apparently, the vector fields $W_\mu^{1,2}$ will have a mass $M_W = \frac{1}{2}\sqrt{2}gF$, whereas the *linear combination*

$$Z_\mu = \frac{g' B_\mu - g W_\mu^3}{\sqrt{g^2 + g'^2}} = \sin\theta_W B_\mu - \cos\theta_W W_\mu^3, \quad \tan\theta_W = \frac{g'}{g}, \qquad (21.5)$$

receives a mass $m_Z = \frac{1}{2}\sqrt{2}F(g^2 + g'^2)^{\frac{1}{2}} = M_W/\cos\theta_W$. The linear combination perpendicular to Z_μ,

$$A_\mu^{\text{em}} = \cos\theta_W B_\mu + \sin\theta_W W_\mu^3, \qquad (21.6)$$

remains massless. This gauge field defines a U(1) subgroup of SU(2)×U(1) that leaves ϕ_0 invariant. This U(1) subgroup is a combination of the U(1) subgroup of SU(2) generated by $\exp(i\chi\sigma_3)$ and the phase rotations $\exp(i\chi)$ associated with the explicit U(1) group with B_μ as its gauge field. It is trivial to verify that the product of these group elements, $\exp(i\chi)\exp(i\chi\sigma_3)$, indeed leaves ϕ_0

invariant. The gauge symmetry associated with this so-called diagonal U(1) subgroup therefore remains *unbroken* and corresponds to electromagnetism.

The Higgs field has three massless components, Re φ_1, Im φ_1 and Imφ_2, all eaten by the vector particles W and Z, and one massive component $\eta \equiv$ Re φ_2 with a mass

$$m_\eta = \sqrt{-2\kappa} = \sqrt{\lambda}F. \tag{21.7}$$

It is this component that is called the Higgs field. It does *not* couple to A_μ^{em}, because like φ_0, also η is not affected by the transformation $\exp(i\chi)\exp(i\chi\sigma_3)$. Alternatively, this can be seen from the covariant derivative

$$D_\mu \begin{pmatrix} \varphi_1 \\ \varphi_2 \end{pmatrix} = \left\{ \partial_\mu - \frac{ig}{2}(\sigma_1 W_\mu^1 + \sigma_2 W_\mu^2) + \frac{ig}{2\cos\theta_W}(\cos^2\theta_W\sigma_3 - \sin^2\theta_W)Z_\mu \right.$$

$$\left. - \frac{ig}{2}\sin\theta_W(\sigma_3 + 1)A_\mu^{\text{em}} \right\} \begin{pmatrix} \varphi_1 \\ \varphi_2 \end{pmatrix}. \tag{21.8}$$

Using the fact that

$$\sigma_3 + 1 = \begin{pmatrix} 2 & 0 \\ 0 & 0 \end{pmatrix}, \tag{21.9}$$

it is clear that φ_2 has no electric charge, whereas φ_1 has a charge $q = -g\sin\theta_W$. As these are would-be Goldstone bosons, 'eaten' by the vector fields, it will turn out that the combinations $W_\mu^\pm = \frac{1}{2}\sqrt{2}(W_\mu^1 \mp iW_\mu^2)$ are charged with an electric charge of $\pm e$, where

$$e = -g\sin\theta_W. \tag{21.10}$$

As a consequence, the two coupling constants g and g' are determined by the electric charge e and the so-called weak mixing angle θ_W, also called the Weinberg angle. From experiment it follows that $\sin^2\theta_W \approx 0.23$. The Z vector field will remain neutral under the electromagnetic interactions. To verify the charge assignment to the vector fields, we have to find the coupling of the various fields to A_μ^{em}. For this it is sufficient to consider the following part of the Lagrangian:

$$\mathcal{L}_{W,B} = -\tfrac{1}{4}\left(\partial_\mu W_\nu^a - \partial_\nu W_\mu^a + g\varepsilon_{abc}W_\mu^b W_\nu^c\right)^2 - \tfrac{1}{4}\left(\partial_\mu B_\nu - \partial_\nu B_\mu\right)^2, \tag{21.11}$$

with the obvious shorthand notations like $(F_{\mu\nu}^a)^2 = F_{\mu\nu}^a F_a^{\mu\nu}$. After some algebra the above equation can be rewritten as

$$\mathcal{L}_{W,B} = -\tfrac{1}{4}\left(F_{\mu\nu}^{\text{em}} + ie(W_\mu^+ W_\nu^- - W_\nu^+ W_\mu^-)\right)^2$$

$$-\tfrac{1}{4}\left(\partial_\mu Z_\nu - \partial_\nu Z_\mu + ig\cos\theta_W(W_\mu^+ W_\nu^- - W_\nu^+ W_\mu^-)\right)^2$$

$$-\tfrac{1}{2}|D_\mu^{\text{em}} W_\nu^- - D_\nu^{\text{em}} W_\mu^- - ig\cos\theta_W(Z_\mu W_\nu^- - Z_\nu W_\mu^-)|^2, \tag{21.12}$$

where we have defined

$$D_\mu^{\text{em}} = \partial_\mu - ie A_\mu^{\text{em}}, \quad F_{\mu\nu}^{\text{em}} = \partial_\mu A_\nu^{\text{em}} - \partial_\nu A_\mu^{\text{em}}. \tag{21.13}$$

We immediately read-off that our charge assignments for Z and $W^\pm$ have been correct. Note that the vector field $W_\mu^\pm$ has an extra magnetic moment because of its coupling to $F_{\mu\nu}^{\text{em}}$

$$\mathcal{L}_{\text{magn.mom.}} = -ie F_{\mu\nu}^{\text{em}} W_\mu^+ W_\nu^-, \tag{21.14}$$

which is a direct consequence of the spin of the vector field. (The magnetic moment for the Dirac field is discussed in Problem 32.)

We now introduce the fermions in the standard model. They are arranged according to *families*. The first family with the smallest masses consists of the electron, the neutrino, the up quark and the down quark. Essential in the standard model is that invariance under parity is broken explicitly by the weak interactions (as has been observed in the beta decay of cobalt-60; see Problem 40). This is achieved by coupling the left- and right-handed helicity eigenstates of the fermions differently to the gauge fields. It should be stressed that the standard model does not explain why parity is violated; it was put in 'by hand.' For each fermion we define

$$\Psi^L = \tfrac{1}{2}(1 - \gamma_5)\Psi, \quad \Psi^R = \tfrac{1}{2}(1 + \gamma_5)\Psi. \tag{21.15}$$

The Dirac Lagrangian in terms of these helicity eigenstates can be written as

$$\mathcal{L}_\Psi = \overline{\Psi}(i\gamma^\mu \partial_\mu - m)\Psi = \overline{\Psi}^R(i\gamma^\mu \partial_\mu)\Psi^R + \overline{\Psi}^L(i\gamma^\mu \partial_\mu)\Psi^L$$
$$-m(\overline{\Psi}^R \Psi^L + \overline{\Psi}^L \Psi^R), \tag{21.16}$$

such that different transformation rules for $\Psi^{R,L}$ enforce $m = 0$, i.e., the absence of an explicit mass term. The beauty of the Higgs mechanism is that it also provides a mass for the fermions. This is achieved by coupling the scalar field ϕ to the fermions, using a Yukawa coupling

$$\mathcal{L}_{\Psi,\phi} = -y(\overline{\Psi}^R \phi^\dagger \Psi^L + \overline{\Psi}^L \phi \Psi^R), \tag{21.17}$$

where y is the Yukawa coupling constant. It also immediately fixes the representation to which $\Psi^{R,L}$ should belong. Since the Lagrangian has to be invariant with respect to the gauge symmetries, and since the scalar field is in the fundamental representation of SU(2), we require that Ψ^L is also in the fundamental representation, i.e., it is a doublet. On the other hand Ψ^R is taken to be invariant under SU(2) (also called the singlet representation). The couplings of the fermions to the gauge field B_μ have to be chosen such that the Lagrangian is neutral. This coupling is parametrised by the so-called hypercharge Y, in units of $-\tfrac{1}{2}g'$.

$$Y_H \equiv Y(\phi) = 1, \quad Y_R = Y_L - 1, \quad Y_{R,L} \equiv Y(\Psi^{R,L}). \tag{21.18}$$

The mass of the fermions is now read-off from eq. (21.17) by replacing ϕ with its so-called vacuum expectation value ϕ_0

$$\mathcal{L}_{\Psi,\phi} = -yF\left(\overline{\Psi}^R\Psi_2^L + \overline{\Psi}_2^L\Psi^R\right), \tag{21.19}$$

where the index on Ψ^L indicates the so-called isospin index, the spinor index of the two-dimensional fundamental representation for the internal SU(2) symmetry group. We also see that Ψ_1^L remains massless and this is exactly the neutrino. The electron is identified with the pair (Ψ^R, Ψ_2^L) and has a mass $m_e = yF$. We want the neutrino to have no electric charge and this fixes the hypercharge of Ψ^L. It is most easily determined from the covariant derivative, acting on the left-handed fermion, defined as in eqs. (21.1) and (21.8), since both are in the same representation. (Electron and neutrino are also neutral with respect to the strong interactions; the situation for the quarks will be discussed below.)

$$D_\mu\Psi^L = \left\{\partial_\mu - \frac{ig}{2}\left(\sigma_1 W_\mu^1 + \sigma_2 W_\mu^2\right) + \frac{ig}{2\cos\theta_W}\left(\cos^2\theta_W\sigma_3 - Y_L\sin^2\theta_W\right)Z_\mu\right.$$
$$\left. - \frac{ig\sin\theta_W}{2}(\sigma_3 + Y_L)A_\mu^{\text{em}}\right\}\Psi^L. \tag{21.20}$$

To make Ψ_1^L decouple from the electromagnetic field, we require

$$Y_L = -1, \quad Y_R = -2. \tag{21.21}$$

This also allows us to find the electric charge of Ψ_2^L to be $g\sin\theta_W = -e$, which as it should be, coincides with the electron charge. The right-handed component should of course have the same electric charge. In that case the covariant derivative is given by

$$D_\mu\Psi^R = \left(\partial_\mu - \frac{ig'}{2}Y_R B_\mu\right)\Psi^R = \left(\partial_\mu - \frac{ig'}{2}Y_R[\sin\theta_W Z_\mu + \cos\theta_W A_\mu^{\text{em}}]\right)\Psi^R$$
$$= \left(\partial_\mu - ie A_\mu^{\text{em}} - ie\tan\theta_W Z_\mu\right)\Psi^R, \tag{21.22}$$

with the expected coupling to the electromagnetic field. Note that we can summarise our assignments of the electric charge by introducing the charge operator in terms of the hypercharge and the so-called isospin operator I_3

$$Q^{\text{em}} = (\tfrac{1}{2}Y + I_3)e. \tag{21.23}$$

On a doublet (Ψ^L and φ) one has $I_3 = \tfrac{1}{2}\sigma_3$, whereas $I_3 = 0$ on a singlet (Ψ^R).

We now discuss quarks. There the weak interactions also act differently on the left- and right-handed components. The left-handed up and down quarks are combined in a doublet representation for SU(2). If we denote the quark fields by q, we assign q_1^L to the left-handed component of the up quark (also denoted by u^L) and q_2^L to the left-handed component of the down quark (d^L).

This doublet gets a hypercharge $Y(q^L) = \frac{1}{3}$, from which we read off the electric charges

$$Q^{em}q^L = Q^{em}\begin{pmatrix} u^L \\ d^L \end{pmatrix} = \begin{pmatrix} \frac{2e}{3}u^L \\ \frac{-e}{3}d^L \end{pmatrix}. \tag{21.24}$$

The right-handed components of both the up and down quarks are singlets under SU(2) and their hypercharges are chosen to ensure that they have the same electric charge as for their left-handed partners

$$Y(u^R) = \frac{4}{3} \quad \text{and} \quad Y(d^R) = -\frac{2}{3}. \tag{21.25}$$

The quarks transform nontrivially under SU(3), the gauge group of the strong interactions. They form complex vectors in the three-dimensional defining or fundamental representation of SU(3). The generators for SU(3) are given by

$$T^1 = -\frac{i}{2}\begin{pmatrix} 0 & 1 & 0 \\ 1 & 0 & 0 \\ 0 & 0 & 0 \end{pmatrix}, \quad T^2 = -\frac{i}{2}\begin{pmatrix} 0 & -i & 0 \\ i & 0 & 0 \\ 0 & 0 & 0 \end{pmatrix}, \quad T^3 = -\frac{i}{2}\begin{pmatrix} 1 & 0 & 0 \\ 0 & -1 & 0 \\ 0 & 0 & 0 \end{pmatrix},$$

$$T^4 = -\frac{i}{2}\begin{pmatrix} 0 & 0 & 1 \\ 0 & 0 & 0 \\ 1 & 0 & 0 \end{pmatrix}, \quad T^5 = -\frac{i}{2}\begin{pmatrix} 0 & 0 & -i \\ 0 & 0 & 0 \\ i & 0 & 0 \end{pmatrix}, \quad T^6 = -\frac{i}{2}\begin{pmatrix} 0 & 0 & 0 \\ 0 & 0 & 1 \\ 0 & 1 & 0 \end{pmatrix},$$

$$T^7 = -\frac{i}{2}\begin{pmatrix} 0 & 0 & 0 \\ 0 & 0 & -i \\ 0 & i & 0 \end{pmatrix}, \quad T^8 = -\frac{i}{2\sqrt{3}}\begin{pmatrix} 1 & 0 & 0 \\ 0 & 1 & 0 \\ 0 & 0 & -2 \end{pmatrix}, \tag{21.26}$$

normalised in accordance with eq. (18.20). (In terms of the so-called Gell–Mann matrices, one has $T^a = -i\lambda_a/2$.) We leave it as an exercise to determine the structure constants. Note that the Lie algebra for the group SU(N) is given by traceless and antihermitian ($X^\dagger = -X$) complex $N \times N$ matrices. The dimension of this Lie algebra is easily seen to be $N^2 - 1$. Note that $\det[\exp(X)] = \exp[\text{Tr}(X)]$, such that $\exp(X)$ has determinant one. Also, $\exp(X)^{-1} = \exp(-X) = \exp(X^\dagger) = \exp(X)^\dagger$ guarantees that $\exp(X)$ is a unitary matrix.

The *fractional* electric charge of the quarks is not observable (otherwise we would have had a different unit for electric charge). The reason is that quarks are conjectured to always form bound states that are neutral under SU(3). This can be achieved by either taking three quarks in an antisymmetric combination to form a SU(3) singlet or by combining a quark and an antiquark. In the first case one has a baryon, of which the proton (*uud*) and the neutron

(udd) are examples. The quark–antiquark bound state is called a meson, of which the pions are examples (e.g., $\pi^+ = u\bar{d}$ and $\pi^- = \bar{u}d$). The bar over the symbol of a particle of course denotes the antiparticle. Rather prosaically one associates to the three SU(3) components of the quark field the property colour. Choosing the three basic colours red, blue, and green makes a bound state of three quarks in an antisymmetric wave function, where hence all colours are different, into a colourless composite. Similarly, combining a quark and an antiquark gives a colourless combination. It is not too difficult to show that a bound state of quarks and antiquarks is a singlet under SU(3) if and only if the net colour is white. It is now also easily verified that with the particular fractional electric charges assigned to the quarks, a colourless combination always has an electric charge that is an integer multiple of the electron charge. For this note that both quarks have modulo e, an electric charge equal to $-\frac{1}{3}e$, whereas both antiquarks have modulo e charge of $\frac{1}{3}e$. Three quarks bound together therefore have zero charge modulo e. The same holds for a quark–antiquark bound state.

That the strong interactions really are strong follows from the fact that a quark and antiquark cannot be separated without creating a quark–antiquark pair from the vacuum, to make sure that the separated components remain neutral under SU(3). This is achieved by combining the quark (antiquark) of the pair created with the antiquark (quark) we try to separate. The mechanism that prevents free quarks from appearing is called confinement, which still lacks a solid theoretical understanding. Because the coupling constant is strong, a perturbative expansion is no longer applicable. That nevertheless the theory of the strong interactions is believed to be the correct theory to describe the forces amongst the quarks (and therefore indirectly the nuclear forces) follows from the remarkable property that at high energies the effective coupling constant is small, and at infinite energy even zero. This is called asymptotic freedom and will only briefly be discussed in the next chapter. For a more detailed discussion we refer to Itzykson and Zuber. In Table 21.1 we list the gauge particles of the standard model.

The strong interactions do not break parity invariance; i.e., the eight gluons A_μ^a couple to the left-handed and right-handed components of the quark fields in the same way. However, the so-called Cabibbo–Kobayashi–Maskawa (CKM) mixing with two other families of quarks (the strange and charm quark on the one hand and the bottom and top quark on the other hand) gives in a very subtle way rise to violation of CP, that is the combination of charge conjugation and parity (equivalent to time reversal T, since CPT is

TABLE 21.1

Gauge particles.

	Name	Charge	Spin	Mass	Force
γ	photon	0	1	0	electromagnetism
A	gluon	0	1	0	strong force
$W^\pm$	W particle	$\pm e$	1	80 GeV	weak force
Z	Z particle	0	1	91 GeV	weak force

TABLE 21.2

Fermion families.

	Name	Charge	Spin	Mass
d	down quark	$-e/3$	1/2	10 MeV
u	up quark	$2e/3$	1/2	5 MeV
e	electron	$-e$	1/2	511 keV
ν_e	neutrino	0	1/2	0(<10 eV)
s	strange quark	$-e/3$	1/2	250 MeV
c	charm quark	$2e/3$	1/2	1.5 GeV
μ	muon	$-e$	1/2	106 MeV
ν_μ	muon-neutrino	0	1/2	0(<0.5 MeV)
b	bottom quark	$-e/3$	1/2	4.8 GeV
t	top quark	$2e/3$	1/2	174 GeV
τ	tau	$-e$	1/2	1.8 GeV
ν_τ	tau-neutrino	0	1/2	0(<164 MeV)

conserved). The electron and neutrino, called leptons, in the first family are replaced by the muon and its neutrino for the second family and by the tau and associated neutrino for the third family. The experiments described in the introduction (see Problem 37) have shown that there are no more than three of these families with massless neutrinos. In the standard model there is room to add a right-handed partner for the neutrino field, which couples to none of the gauge fields. With a suitably chosen Yukawa coupling, the neutrino can be given an arbitrarily small mass. It is experimentally very hard to measure the mass of the neutrino; only upper bounds have been established. Table 21.2 lists the properties of all the fermions observed in the standard model (the top quark was only discovered in 1994 at Fermilab). For much more on the standard model, see in particular the book by J.C. Taylor mentioned in the introduction.

22

Loop Corrections and Renormalisation

DOI: 10.1201/b15364-22

Up to now, we have only considered the lowest-order calculations of cross sections, for which it is sufficient to consider tree-level diagrams that do not contain any loops. Loop integrals typically give rise to infinities, which can be regularised by considering, for example, a cutoff in momentum space, as was discussed in Chapter 7. Another possibility of regularising the theory is by discretising space-time, amounting to a lattice formulation; see Equation (7.5). In both of these cases there exists a maximal energy (equivalent to a minimal distance). The parameters, like the coupling constants, masses and field renormalisation constants, will depend on this cutoff parameter, generically denoted by an energy Λ or a distance $a = 1/\Lambda$. How to give a physical definition of the mass in terms of the full propagator and why field renormalisation is necessary was discussed in Chapter 9. For the renormalisation of the coupling constant, it is best to define the physical coupling constant in terms of a particular scattering process, as that is what can be measured in experiment. Alternatively, as these are strongly related, the physical couplings can be defined in terms of an amputated $1PI$ n-point function, with prescribed momenta assigned to the external lines, all proportional to an energy scale called $\mu \ll \Lambda$. As an example consider the self-interacting scalar field, with a four-point coupling λ [see, for example, Equation (21.2)]. We define the physical four-point coupling constant in terms of the $1PI$ four-point function with the momenta on the amputated lines set to some particular value proportional to μ (the precise choice is not important for the present discussion). It is clear that this gives a function $\lambda_{\text{reg}}(\lambda, \mu, \Lambda)$. The dependence on other coupling constants and the mass parameters is left implicit.

The theory is considered renormalisable if we can remove the cutoff by adjusting λ (also called the bare coupling constant) in such a way that at a fixed value $\mu = \mu_0$ the renormalised coupling λ_R stays finite and takes on a prescribed (i.e., measured) value. It is then obvious that the renormalised coupling constant $\lambda_R(\mu) \equiv \lambda_{\text{reg}}[\lambda(\Lambda), \mu, \Lambda]$ is a function of μ, coinciding at μ_0 with the prescribed value λ_R. Since the physical coupling constant is computed in terms of the full $1PI$ four-point function, the dependence on the energy scale is caused by quantum corrections. Since the vacuum in field theories is not really empty, as was discussed in the context of the Casimir effect in Chapter 2, the computation is not much different from calculating effective interactions in a polarised medium. In this case the polarisation is due to the virtual particles that describe the quantum fluctuations (zero-point fluctuations) of the vacuum, and is therefore also called the vacuum polarisation. The energy-dependent couplings are called running couplings. It should be emphasised that the running of the couplings is a manifestation of an anomaly (called the conformal anomaly), which is the breaking of a symmetry of the Lagrangian by the quantum corrections. In the absence of a mass, the scalar field theory with a φ^4 interaction is at the classical level invariant under scale transformations, $\varphi(x) \rightarrow \kappa\varphi(x/\kappa)$, where κ is the scale parameter. It is obvious that the regularised couplings are not invariant under such a rescaling because of the presence of a cutoff. What is not obvious is that, for the simple field theories we have been considering in four dimensions, the scale independence cannot be recovered by removing the cutoff (i.e., taking $\Lambda \rightarrow \infty$).

By adjusting the bare coupling constants of the theory as a function of the cutoff Λ, to ensure that all regularised couplings stay finite when the cutoff is moved to infinity, the calculations can be arranged such that nowhere explicit infinities occur. When we say that the contributions of the loops diverge, we mean that without adjusting the bare coupling constant, their contributions are infinite in the limit $\Lambda \rightarrow \infty$. A theory is called renormalisable if only a finite number of bare coupling constants needs to be adjusted to have all $1PI$ n-point functions finite. This can be shown to be equivalent to all $1PI$ n-point functions to be completely determined as a function of a finite number of renormalised couplings, called relevant couplings. It is only in such an instance that quantum field theory has predictive power. Renormalisability is therefore a necessary requirement for the theory to be insensitive to what happens at very high energies with a maximal amount of predictive power. The standard model falls in this class of theories.

Theoretical studies of the last five years or so have shown that the self-coupling of the Higgs field will most likely vanish if we really take $\Lambda \rightarrow \infty$, albeit in a logarithmic way. Loosely speaking the running of this coupling is such that the renormalised coupling increases with increasing energy. The only way it can be avoided (that the renormalised coupling will become infinite at some finite energy) is to either take the renormalised coupling equal to zero or to keep the cutoff finite. It depends on the parameters of the model, in particular the Higgs mass, how large the cutoff should be. If the Higgs mass is relatively light, this can be at the Planck scale and has little consequence for the theory. If, however, the Higgs mass turns out to be in the order of 1 TeV, the cutoff has to be roughly smaller than 10 TeV.

As we can measure the self-coupling of the Higgs field and related quantities to be nontrivial (which is, of course, crucial for the spontaneous symmetry breaking and giving a mass to the W and Z particles), the scalar sector of the standard model depends in a rather subtle way on what happens at higher energies. This sensitivity to high energies is, however, much weaker than in nonrenormalisable theories like for the four-fermi interactions. It is outside the scope of these lectures to describe the computations necessary to make the above more precise. In the following we give a sample calculation that will provide the technical ingredients to perform such calculations and to illustrate some of these issues in a simple setting. Also, Problems 2, 38, and 39 illustrate further ingredients that are pertinent to renormalising field theories.

Let us end this discussion by noting that a running coupling can, of course, either increase or decrease at increasing energy. The Higgs self-coupling and the electromagnetic coupling constant e are examples of couplings that increase at high energies. For the electric charge e, this increase is very tiny and the cutoff can be chosen much bigger than the Planck energy. The analogy with a polarised medium is that the virtual particles in the field of a charged particle will screen its charge at large distances. When we probe the charged particle at ever smaller distances, the effective charge becomes less screened and increases. Due to the self-interactions of a non-Abelian gauge field, its charges show the effect of antiscreening. Here the effective charge becomes bigger at larger distances. For the strong interactions this is one way to understand confinement. The energy of a single quark within a spherical shell would increase without bound with increasing radius. A free quark would carry an infinite energy. To the contrary, at decreasing separations, the effective charge becomes weaker and weaker and the quarks start to behave as free particles. This is the asymptotic freedom mentioned in the previous chapter.

We will now consider to one-loop order the self-energy for the scalar field φ with a mass m, coupled to two flavours of fermions with masses m_1 and m_2, coupled through Yukawa couplings described by the Lagrangian

$$\mathcal{L} = \tfrac{1}{2}(\partial_\mu \varphi)^2 - \tfrac{1}{2}m^2\varphi^2 - \tfrac{1}{6}\lambda\varphi^3 + \sum_i \overline{\Psi}_{(i)}(i\gamma^\mu \partial_\mu - m_i)\Psi_{(i)}$$

$$-g\varphi\left(\overline{\Psi}_{(1)}\Psi_{(2)} + \overline{\Psi}_{(2)}\Psi_{(1)}\right). \tag{22.1}$$

The self-energy for the scalar field in one-loop order splits in two contributions Σ_1 and Σ_2 from a fermion and a scalar loop (in this order $Z \equiv 1$):

$$\Sigma_\varphi(q) \equiv \Sigma_1 + \Sigma_2 = - \quad + \quad . \tag{22.2}$$

The numerical expressions for Σ_1 and Σ_2 are given by

$$\Sigma_1 = -\frac{ig^2}{(2\pi)^4}\int d_4 k \, \frac{\text{Tr}((\not{k}+m_1)(\not{k}+\not{q}+m_2))}{(k^2 - m_1^2 + i\varepsilon)((k+q)^2 - m_2^2 + i\varepsilon)},$$

$$\Sigma_2 = \frac{i\lambda^2}{2(2\pi)^4}\int d_4 k \, \frac{1}{(k^2 - m^2 + i\varepsilon)((k+q)^2 - m^2 + i\varepsilon)}. \tag{22.3}$$

Using Equation (11.1), requires us to employ the Feynman rules of Table 9.1 to obtain these expressions. [Alternatively the Feynman rules of Table 8.1 can be used, provided the self-energy is defined through Equation (9.9).] These integrals are obviously divergent. Introducing a momentum cutoff Λ we find $\Sigma_1 \sim \Lambda^2$ and $\Sigma_2 \sim \log \Lambda$ to lowest nontrivial order in $1/\Lambda$. One says that Σ_1

is quadratically and Σ_2 logarithmically divergent. To simplify the integrands we discuss the Feynman trick

$$\frac{1}{ab} = \int_0^1 dx \, \frac{1}{\left(ax + b(1-x)\right)^2}, \tag{22.4}$$

which we can apply to the computation of Σ_1 by substituting $a = (k+q)^2 - m_2^2 + i\varepsilon$ and $b = k^2 - m_1^2 + i\varepsilon$. For Σ_2 we have the same assignment, with in addition $m_1 = m_2 = m$. It is useful to also have the generalisation of the Feynman trick for an arbitrary product of scalar propagators,

$$\left(\prod_{i=1}^k a_i^{n_i}\right)^{-1} = \frac{\Gamma\left(\sum_i n_i\right)}{\prod_{i=1}^k \Gamma(n_i)} \prod_{i=1}^k \left(\int_0^1 x_i^{n_i-1} dx_i\right)$$

$$\times \delta\left(\sum_{i=1}^k x_i - 1\right)\left(\sum_{i=1}^k a_i x_i\right)^{-\sum_i n_i}, \tag{22.5}$$

which is proven by induction. In here $\Gamma(z)$ is the gamma function, which satisfies $\Gamma(z+1) = z\Gamma(z)$, $\Gamma(n+1) = n!$ and $\Gamma(\frac{1}{2}) = \sqrt{\pi}$ [see Problem 2(b)]. Consequently we find

$$\Sigma_1 = -\frac{4ig^2}{(2\pi)^4} \int d_4k \int_0^1 dx$$

$$\times \frac{k^2 + k \cdot q + m_1 m_2}{\left((k + (1-x)q)^2 + x(1-x)q^2 - xm_1^2 - (1-x)m_2^2 + i\varepsilon\right)^2},$$

$$\Sigma_2 = \frac{i\lambda^2}{2(2\pi)^4} \int d_4k \int_0^1 dx \frac{1}{\left((k + (1-x)q)^2 + x(1-x)q^2 - m^2 + i\varepsilon\right)^2}. \tag{22.6}$$

We will show how to regularise these two integrals in two different ways. First we use dimensional regularisation introduced by 't Hooft and Veltman (see Problem 2) and then discuss Pauli–Villars regularisation. In dimensional regularisation the loop integrations are replaced by integrals in n, instead of four, dimensions. The momentum integrations are always of the form

$$I_{n,\alpha,\beta}(M) \equiv \int d_nk \, \frac{(k^2)^\alpha}{(k^2 - M^2 + i\varepsilon)^\beta}. \tag{22.7}$$

We can evaluate this integral by performing the so-called Wick rotation, where we replace the integral over $\mathrm{Re}k_0$ by an integration over $\mathrm{Im}k_0$. The integral over the two quarter circles indicated in Figure 22.1 will vanish as the radius tends to infinity. As there are no poles inside the contour of integration, we find

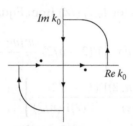

FIGURE 22.1
Wick rotation.

$$I_{n,\alpha,\beta}(M) = i(-1)^{\alpha-\beta} \int d_n k \, \frac{(k^2)^\alpha}{(k^2 + M^2 - i\varepsilon)^\beta}, \quad k^2 = k_0^2 + \vec{k}^2. \quad (22.8)$$

We note that the integrand is a purely radial integral and as the surface area of an n-dimensional sphere is analytically known [$S_n = 2\pi^{n/2}/\Gamma(n/2)$, e.g., $S_2 = 2\pi$, $S_3 = 4\pi$, $S_4 = 2\pi^2$,...], we obtain

$$I_{n,\alpha,\beta}(M) = i(-1)^{\alpha-\beta} \frac{2\pi^{n/2}}{\Gamma(n/2)} \int r^{n-1} dr \, \frac{r^{2\alpha}}{(r^2 + M^2 - i\varepsilon)^\beta}$$

$$= \frac{i(-1)^{\alpha-\beta}\pi^{n/2}\Gamma(\alpha+n/2)\Gamma(\beta-\alpha-n/2)}{(M^2 - i\varepsilon)^{\beta-\alpha-n/2}\Gamma(\beta)\Gamma(n/2)}. \quad (22.9)$$

We used the integral representation of the beta function

$$B(n,k) \equiv \frac{\Gamma(n)\Gamma(k)}{\Gamma(n+k)} = \int_0^\infty dz \, \frac{z^{n-1}}{(z+1)^{k+n}}. \quad (22.10)$$

Shifting the integration variable $k \to k - (1-x)q$, we find

$$\Sigma_1 = -\frac{4ig^2}{(2\pi)^4} \int d_n k \int_0^1 dx \, \frac{k^2 - (1-2x)k \cdot q + m_1 m_2 + x(x-1)q^2}{(k^2 - \hat{M}_{x,q}^2 + i\varepsilon)^2},$$

$$\Sigma_2 = \frac{i\lambda^2}{2(2\pi)^4} \int d_n k \int_0^1 dx \, \frac{1}{(k^2 - M_{x,q}^2 + i\varepsilon)^2}, \quad (22.11)$$

where

$$\hat{M}_{x,q}^2 = xm_1^2 + (1-x)m_2^2 - x(1-x)q^2 \quad \text{and} \quad M_{x,q}^2 = m^2 - x(1-x)q^2, \quad (22.12)$$

which allows us to express Σ_i in terms of the integrals $I_{n,\alpha,\beta}(m)$

$$\Sigma_1 = -\frac{4ig^2}{(2\pi)^4} \int_0^1 dx \left\{ I_{n,1,2}(\hat{M}_{x,q}) + (m_1 m_2 - x(1-x)q^2)I_{n,0,2}(\hat{M}_{x,q}) \right\},$$

$$\Sigma_2 = \frac{i\lambda^2}{2(2\pi)^4} \int_0^1 dx \, I_{n,0,2}(M_{x,q}). \quad (22.13)$$

Substituting the expressions for $I_{n,\alpha,\beta}(M)$ from Equation (22.9), we find

$$
\Sigma_1 = \frac{4g^2\pi^{n/2}}{(2\pi)^4} \int_0^1 dx \left\{ \Gamma(2-n/2) \frac{(m_1 m_2 - x(1-x)q^2)}{(\hat{M}_{x,q}^2 - i\varepsilon)^{2-n/2}\Gamma(2)} \right.
$$
$$
\left. - \frac{\Gamma(1+n/2)\Gamma(1-n/2)}{(\hat{M}_{x,q}^2 - i\varepsilon)^{1-n/2}\Gamma(2)\Gamma(n/2)} \right\},
$$
$$
\Sigma_2 = -\frac{\lambda^2\pi^{n/2}}{2(2\pi)^4} \int_0^1 dx \, \frac{\Gamma(2-n/2)}{(M_{x,q}^2 - i\varepsilon)^{2-n/2}\Gamma(2)}. \tag{22.14}
$$

This can be further simplified using

$$
\frac{\Gamma(1+n/2)\Gamma(1-n/2)}{\Gamma(2-n/2)\Gamma(n/2)} = \frac{n}{2-n}, \tag{22.15}
$$

such that

$$
\Sigma_1 = \frac{4g^2\pi^{n/2}\Gamma(2-n/2)}{(2\pi)^4} \int_0^1 dx \left\{ \frac{(m_1 m_2 - x(1-x)q^2)}{(\hat{M}_{x,q}^2 - i\varepsilon)^{2-n/2}} - \frac{n/(2-n)}{(\hat{M}_{x,q}^2 - i\varepsilon)^{1-n/2}} \right\},
$$
$$
\Sigma_2 = -\frac{\lambda^2\pi^{n/2}\Gamma(2-n/2)}{2(2\pi)^4} \int_0^1 dx \, \frac{1}{(M_{x,q}^2 - i\varepsilon)^{2-n/2}}. \tag{22.16}
$$

The divergent part is now fully contained in $\Gamma(2-n/2)$, because

$$
\Gamma(2-n/2) = \frac{\Gamma\left(1+\frac{1}{2}(4-n)\right)}{\frac{1}{2}(4-n)} = \frac{2}{4-n} - \gamma + \mathcal{O}(4-n), \tag{22.17}
$$

where $\gamma = 0.57721\ldots$ is the Euler constant. We expand Σ_i around $n=4$

$$
\Sigma_1 = \frac{g^2}{(2\pi)^2}\left(\frac{2}{4-n}-\gamma\right)\pi^{(n-4)/2}\int_0^1 dx
$$
$$
\times \left(\frac{m_1 m_2 - x(1-x)q^2 + (2+\frac{1}{2}(4-n))\hat{M}_{x,q}^2}{(\hat{M}_{x,q}^2 - i\varepsilon)^{(4-n)/2}}\right)
$$
$$
= \frac{g^2}{(2\pi)^2}\left(\frac{2}{4-n}-\gamma\right)\int_0^1 dx \left\{ (m_1 m_2 - 3x(1-x)q^2 + 2xm_1^2 + 2(1-x)m_2^2) \right.
$$
$$
\left. \times \left[1 - \tfrac{1}{2}(4-n)\log\left(\pi[\hat{M}_{x,q}^2 - i\varepsilon]\right)\right] + \tfrac{1}{2}(4-n)\hat{M}_{x,q}^2 \right\}
$$
$$
\equiv \frac{\Sigma_1^{(-1)}}{4-n} + \Sigma_1^{(0)} + \mathcal{O}(4-n),
$$
$$
\Sigma_2 = \frac{\lambda^2}{8(2\pi)^2}\left\{\int_0^1 dx \,\left(\log\left(\pi[M_{x,q}^2 - i\varepsilon]\right) + \gamma\right) - \frac{2}{4-n}\right\}
$$
$$
\equiv \frac{\Sigma_2^{(-1)}}{4-n} + \Sigma_2^{(0)} + \mathcal{O}(4-n). \tag{22.18}
$$

Note that $q^2 = q_0^2 - \vec{q}^{\,2}$ and that the coupling constant λ for the scalar three-point coupling has the dimension of mass. We have split the result for Σ_i in a pole term with residue $\Sigma_i^{(-1)}$ and a finite part $\Sigma_i^{(0)}$ for $n \to 4$.

$$\Sigma_1^{(0)} = \frac{g^2}{4\pi^2} \int_0^1 dx \left(\hat{M}_{x,q}^2 - \left\{ \log\left(\pi \left[\hat{M}_{x,q}^2 - i\varepsilon \right] \right) + \gamma \right\} \right.$$

$$\left. \times \left(m_1 m_2 + 3x(x-1)q^2 + 2xm_1^2 + 2(1-x)m_2^2 \right) \right),$$

$$\Sigma_2^{(0)} = \frac{\lambda^2}{8(2\pi)^2} \int_0^1 dx \left(\log\left(\pi \left[M_{x,q}^2 - i\varepsilon \right] \right) + \gamma \right),$$

$$\Sigma_1^{(-1)} = \frac{g^2}{2\pi^2}(m_1 m_2 + m_1^2 + m_2^2 - \tfrac{1}{2}q^2), \quad \Sigma_2^{(-1)} = -\frac{\lambda^2}{(4\pi)^2}. \quad (22.19)$$

We now note that the pole terms are of the same form as the tree-level expressions obtained from the following extra term in the Lagrangian:

$$\Delta \mathcal{L} = \tfrac{1}{2}a(\partial_\mu \varphi)^2 - \tfrac{1}{2}b\varphi^2. \quad (22.20)$$

This means that we can choose a and b so as to precisely cancel the pole terms. To lowest order we therefore have

$$\Sigma_\varphi(\mathcal{L} + \Delta \mathcal{L}) = \Sigma_\varphi(\mathcal{L}) + b - aq^2 = \Sigma_1^{(0)} + \Sigma_2^{(0)} + \mathcal{O}(n-4), \quad (22.21)$$

from which we can solve for a and b in terms of $\Sigma_i^{(-1)}$

$$a = -\frac{g^2}{(2\pi)^2}\frac{1}{4-n}, \quad b = \left(\frac{\lambda^2}{(4\pi)^2} - \frac{g^2}{2\pi^2}(m_1 m_2 + m_1^2 + m_2^2) \right) \frac{1}{4-n}. \quad (22.22)$$

Note that as long as we stay away from $n = 4$ everything is well defined, including a and b. The limit $n \to 4$ is to be taken after we have expressed everything in terms of the renormalised coupling constants and masses. We have taken here a slightly different approach for renormalising the theory. Rather than computing at $n \neq 4$ physical processes to fix the renormalised couplings, we have started with renormalised couplings and determined how they have to depend on the bare couplings so as to cancel any infinities that might arise as $n \to 4$. It is clear that these two procedures are equivalent. For the physical interpretation, the first procedure (due to Wilson) is more transparent; in a loop expansion, the second procedure is more natural. To find the bare mass and the field renormalisation (for the bare λ coupling, we should have considered the $1PI$ three-point function with three φ external lines), we write to one-loop order

$$\mathcal{L}_B = \mathcal{L} + \Delta \mathcal{L} \equiv \tfrac{1}{2}(\partial_\mu \varphi_B)^2 - \tfrac{1}{2}m_B^2 \varphi_B^2, \quad m_B^2 \equiv \frac{m^2 + b}{1+a},$$

$$\varphi_B \equiv \sqrt{Z_\varphi}\varphi = \sqrt{1+a}\,\varphi. \quad (22.23)$$

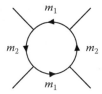

FIGURE 22.2
Four point diagram with fermion loop.

Often it can be determined by power-counting (of momenta) which diagrams give rise to divergencies for $n \rightarrow 4$ or $\Lambda \rightarrow \infty$. The infinities correspond to local counter terms (i.e, with a finite number of space-time derivatives) in the Lagrangian. For the theory in Equation (22.1), power counting easily shows that the φ four-point function is logarithmically divergent at one-loop order; see the Feynman diagram, Figure 22.2. We therefore need to introduce an independent parameter for the φ four-point coupling, so as to adjust its bare coupling to depend in the proper way on the cutoff, to ensure that we can remove it. It can be shown that after adding to the Lagrangian in Equation (22.1) the term $-\lambda_4 \varphi^4 / 4!$, the theory becomes renormalisable to all orders in the loop expansion. The relevant parameters are m, m_1, m_2, g, λ and λ_4.

As we have seen in Chapter 11, $\Sigma_\varphi(q)$ should have a nonvanishing imaginary part if the scalar particle is unstable. It is clear that the scalar particle itself cannot decay in two scalar particles, but when $m_1 + m_2 < m$ it could decay in two fermions. Indeed, it is not difficult to show that on the mass shell $(q^2 = m^2)$ $\Sigma_2^{(0)}$ is real

$$\Sigma_2^{(0)}(q^2 = m^2) = \frac{\lambda^2}{8(2\pi)^2} \int_0^1 dx \, \log[(x - \tfrac{1}{2})^2 + \tfrac{3}{4}] + \gamma + \log(m^2 \pi) \in \mathbb{R}.$$

$$(22.24)$$

We will show that $\mathrm{Im}\Sigma_1^{(0)}(q^2 = m^2) \neq 0$ if and only if $q^2 > (m_1 + m_2)^2$, called the threshold for decay. The only way $\Sigma_1^{(0)}$ can develop an imaginary part is when the argument of the logarithm in Equation (22.19) becomes negative. The threshold is therefore determined by

$$\min\{\hat{M}_{x,q}^2 | x \in [0, 1]\} = \min\{x m_1^2 + (1 - x) m_2^2 - x(1 - x) q^2 | x \in [0, 1]\} < 0.$$

$$(22.25)$$

Let us first consider the simplest case of equal fermion masses, $m_1 = m_2$. In that case

$$\min\{\hat{M}_{x,q}^2 | x \in [0, 1]\} = m_1^2 - \tfrac{1}{4} q^2 \quad (m_1 = m_2),$$

$$(22.26)$$

and the threshold is determined by $q^2 > 4m_1^2 = (m_1 + m_2)^2$. For the general case of unequal fermion masses, the minimum is obtained for $x = \tfrac{1}{2}[1+$

$(m_2^2 - m_1^2)/q^2]$. After some algebra we find

$$\min\{\hat{M}_{x,q}^2 | x \in \mathbb{R}\} = \frac{q^2}{4}\left(1 - \frac{(m_1 - m_2)^2}{q^2}\right)\left(\frac{(m_1 + m_2)^2}{q^2} - 1\right), \quad (22.27)$$

which is indeed negative for $q^2 > (m_1 + m_2)^2$. The value of x where this minimum is attained does not lie in the interval $[0, 1]$ if $|(m_1^2 - m_2^2)/q^2| > 1$, which can be used to rule out the other region, $q^2 < (m_1 - m_2)^2$, where Equation (22.27) is negative. This therefore proves that the kinematically determined threshold coincides with the threshold for $\text{Im}\Sigma(q) \neq 0$, as was assumed in Chapter 11.

A major advantage of dimensional regularisation is that it preserves the Lorentz and gauge invariances. Furthermore, it is a local regulator. The lattice regularisation also can be arranged to preserve the gauge invariance, but locality and Lorentz invariance are only valid at distances much bigger than the lattice spacing a. A momentum cutoff breaks both the Lorentz and gauge invariance. Pauli–Villars regularisation is aimed at having a regulator that preserves the Lorentz invariance. We will describe it for the Lagrangian of Equation (22.1), using again the computation of $\Sigma_\varphi(q)$ to one-loop order as an illustration. For each of the original fields φ and $\Psi_{(i)}$ one adds extra (ghost) fields, with either the same ($e_\ell > 0$) or opposite ($e_\ell < 0$) statistics. This means that a loop of these ghost fields gets an additional factor e_ℓ. Furthermore, the mass of these ghost fields is shifted over M_ℓ with respect to the original ("parent") field. (Alternatively, if the original field is a boson, one can shift m^2 over M_ℓ^2; see Problem 39. With the present prescription we can treat bosons and fermions on the same footing.) By defining $e_0 = 1$ and $M_0 = 0$, the index $\ell = 0$ describes the original fields of the model. We define furthermore $M_\ell \equiv b_\ell \Lambda$ with $\Lambda \gg m, m_1, m_2$. To regularise Σ_φ by Pauli–Villars' method, we choose

$$e_\ell = (1, -1, 2, -2) \quad \text{and} \quad b_\ell = (0, 4, 3, 1); \quad (22.28)$$

in other words the scalar and two fermion fields each has three ghost fields associated to them but with nonstandard weights for the loops. We could stick to standard weights, such that these ghost fields can be described in terms of either Grassmann or bosonic variables by taking $|e_\ell|$ fields, having either the same ($e_\ell > 0$) or reversed ($e_\ell < 0$) statistics with respect to the original ('parent') field. It is straightforward to give the self-energy including the contribution of the ghost fields

$$\Sigma_\varphi^{\text{PV}}(q) = \sum_{\ell=0}^{3} e_\ell \left\{ \Sigma_1(m_1 + b_\ell \Lambda, m_2 + b_\ell \Lambda; q) + \Sigma_2(m + b_\ell \Lambda; q) \right\}, \quad (22.29)$$

in an obvious notation. The weights are chosen such that the momentum integrals can all be performed. Nevertheless, the masses of the ghost particles, all proportional to Λ, now play the role of a momentum cutoff, as the Lagrangian will at that energy scale no longer describe a physical theory. It is still convenient to evaluate the integrals in n, rather than in four dimensions. We will see that Equation (22.28) guarantees that the terms proportional to $(4-n)^{-1}$ exactly cancel. Indeed, using Equation (22.19)

$$
\sum_{\ell=0}^{3} e_\ell \left(\Sigma_1^{(-1)}(m_1 + b_\ell \Lambda, m_2 + b_\ell \Lambda; q) + \Sigma_2^{(-1)}(m + b_\ell \Lambda; q) \right)
$$

$$
= \sum_{\ell=0}^{3} e_\ell \left(\frac{g^2}{2\pi^2} \left((m_1 + b_\ell \Lambda)(m_2 + b_\ell \Lambda) + (m_1 + b_\ell \Lambda)^2 + (m_2 + b_\ell \Lambda)^2 \right) \right.
$$

$$
\left. - \frac{g^2 q^2}{(2\pi)^2} - \frac{\lambda^2}{(4\pi)^2} \right)
$$

$$
= \frac{g^2(m_1 m_2 + m_1^2 + m_2^2 - \frac{1}{2}q^2) - \frac{1}{8}\lambda^2}{2\pi^2} \sum_\ell e_\ell + \frac{3(m_1 + m_2)g^2}{2\pi^2} \Lambda \sum_\ell e_\ell b_\ell
$$

$$
+ \frac{3g^2}{2\pi^2} \Lambda^2 \sum_\ell e_\ell b_\ell^2 = 0. \tag{22.30}
$$

The finite result that remains (replacing $\Sigma^{(-1)}$ in the equation above by $\Sigma^{(0)}$) is nevertheless still dependent on Λ. To keep the following computation transparent we take $m_1 = m_2$

$$
\Sigma_\varphi^{PV}(q) = \frac{g^2}{(2\pi)^2} \int_0^1 dx \sum_\ell e_\ell \left\{ 1 - 3\gamma - 3 \log \left(\pi [(m_1 + b_\ell \Lambda)^2 - x(1-x)q^2] \right) \right\}
$$

$$
\times \left[(m_1 + b_\ell \Lambda)^2 - x(1-x)q^2 \right]
$$

$$
+ \frac{\lambda^2}{8(2\pi)^2} \int_0^1 dx \sum_\ell e_\ell \left\{ \log \left(\pi [(m + b_\ell \Lambda)^2 - x(1-x)q^2] \right) + \gamma \right\}
$$

$$
= -\frac{3g^2}{(2\pi)^2} \int_0^1 dx \sum_\ell e_\ell \log \left((m_1 + b_\ell \Lambda)^2 - x(1-x)q^2 \right)
$$

$$
\times \left[(m_1 + b_\ell \Lambda)^2 - x(1-x)q^2 \right]
$$

$$
+ \frac{\lambda^2}{8(2\pi)^2} \int_0^1 dx \sum_\ell e_\ell \log \left((m + b_\ell \Lambda)^2 - x(1-x)q^2 \right)
$$

$$
= \frac{\lambda^2}{8(2\pi)^2} \int_0^1 dx \, \log \left(m^2 - x(1-x)q^2 \right)
$$

$$
- \frac{3g^2}{(2\pi)^2} \int_0^1 dx \, \log \left(m_1^2 - x(1-x)q^2 \right) \left[m_1^2 - x(1-x)q^2 \right]
$$

$$
+ a_1 \Lambda^2 + a_2 \Lambda + a_3 \log \Lambda + a_4 q^2 \log \Lambda + a_5 + a_6 q^2 + \mathcal{O}(1/\Lambda). \tag{22.31}
$$

The precise values of the coefficients a_i are not very important but can be calculated explicitly with some effort. All Λ-dependent terms can be absorbed in a redefinition of Z_φ and the mass of the scalar field, such that

$$
\begin{aligned}
\Sigma_\varphi^{PV}(q) = &-\frac{3g^2}{(2\pi)^2} \int_0^1 dx \, \log\left(m_1^2 - x(1-x)q^2\right)\left[m_1^2 - x(1-x)q^2\right] \\
&+\frac{\lambda^2}{8(2\pi)^2} \log\left(m^2 - x(1-x)q^2\right) + a_5 + a_6 q^2.
\end{aligned}
\tag{22.32}
$$

Note that in dimensional regularisation (DR) we found the result $\Sigma_1^{(0)} + \Sigma_2^{(0)}$, or

$$
\begin{aligned}
\Sigma_\varphi^{DR}(q) = &-\frac{3g^2}{(2\pi)^2} \int_0^1 dx \left\{\log\left[\pi\left(m_1^2 - x(1-x)q^2\right)\right] + \gamma - \tfrac{1}{3}\right\} \\
&\times \left[m_1^2 - x(1-x)q^2\right] \\
&+\frac{\lambda^2}{8(2\pi)^2} \int_0^1 dx \left\{\log\left[\pi\left(m^2 - x(1-x)q^2\right)\right] + \gamma\right\}.
\end{aligned}
\tag{22.33}
$$

However, the difference $\Sigma_\varphi^{PV}(q) - \Sigma_\varphi^{DR}(q) = a_5 - b_1 + (a_6 - b_2)q^2$, where

$$
b_1 = \frac{\lambda^2(\gamma + \log\pi)}{8(2\pi)^2} - \frac{g^2 m_1^2(3\gamma + 3\log\pi - 1)}{(2\pi)^2},
$$

$$
b_2 = \frac{g^2 m_1^2(3\gamma + 3\log\pi - 1)}{6(2\pi)^2},
\tag{22.34}
$$

can be absorbed in a finite redefinition of the mass and of Z_φ. If we define the renormalised coupling in terms of some physical scattering process, such an ambiguity of course cannot arise. In that case there is a unique relation between the bare and renormalised parameters. This relation, however, depends on the regularisation used.

We will now discuss, without a detailed derivation, the renormalisation of gauge theories to one-loop order in dimensional regularisation. The bare Lagrangian is given by

$$
\begin{aligned}
\mathcal{L} = &-\tfrac{1}{4}(\partial^\mu A_B^\nu - \partial^\nu A_B^\mu)^2 - \tfrac{1}{2}\alpha_B(\partial_\mu A_B^\mu)^2 + \overline{\Psi}_B(i\gamma^\mu \partial_\mu - m_B)\Psi_B \\
&+ e_B A_B^\mu \overline{\Psi}_B \gamma_\mu \Psi_B.
\end{aligned}
\tag{22.35}
$$

In n dimensions we still want the action to be dimensionless ($\hbar = 1$), which implies that $\mathcal{L}/\mu^n$ is dimensionless. From this we derive the dimensions of the fields and the parameters in n dimensions,

$$
[A_B^\mu] = \mu^{\frac{1}{2}n-1}, \quad [\alpha_B] = 1, \quad [\Psi_B] = [\overline{\Psi}_B] = \mu^{\frac{1}{2}n-\frac{1}{2}},
$$

$$
[m_B] = \mu, \quad [e_B] = \mu^{2-\frac{1}{2}n}.
\tag{22.36}
$$

If we define (as is customary) $\varepsilon \equiv 4 - n$, we find (for details see Itzykson and Zuber, e.g., Sections 7-1 and 8-4. They use slightly different notations.):

$$A_B^\mu \equiv \mu^{-\frac{1}{2}\varepsilon} \sqrt{Z_A} A^\mu = \mu^{-\frac{1}{2}\varepsilon} \left(1 - \frac{e^2}{6\pi^2} \cdot \frac{1}{\varepsilon} + \cdots\right)^{\frac{1}{2}} A^\mu,$$

$$m_B \equiv Z_m m = \left(1 - \frac{3e^2}{8\pi^2} \cdot \frac{1}{\varepsilon} + \cdots\right) m,$$

$$\Psi_B \equiv \mu^{-\frac{1}{2}\varepsilon} \sqrt{Z_\Psi} \Psi = \mu^{-\frac{1}{2}\varepsilon} \left(1 - \frac{e^2}{8\pi^2} \cdot \frac{1}{\varepsilon} + \cdots\right)^{\frac{1}{2}} \Psi,$$

$$\alpha_B \equiv Z_\alpha \alpha = \left(1 + \frac{e^2}{6\pi^2} \cdot \frac{1}{\varepsilon} + \cdots\right) \alpha,$$

$$e_B \equiv \mu^{\frac{1}{2}\varepsilon} Z_e e = \mu^{\frac{1}{2}\varepsilon} \left(1 + \frac{e^2}{12\pi^2} \cdot \frac{1}{\varepsilon} + \cdots\right) e. \qquad (22.37)$$

We note that to one-loop order $A^\mu \equiv e_B A_B^\mu$ and α_B/e_B^2 are finite for $n \to 4$. This is not an accident but the consequence of a so-called Ward identity, which as a consequence of the gauge symmetry (through the BRS invariance mentioned at the end of Chapter 20) relates different Z factors,

$$Z_e^2 = Z_\alpha = 1/Z_A. \qquad (22.38)$$

It is therefore sometimes much more convenient to use

$$\mathcal{L} = -\frac{1}{4e_B^2}(\partial^\mu A^\nu - \partial^\nu A^\mu)^2 - \frac{\alpha_B}{2e_B^2}(\partial_\mu A^\mu)^2 + \overline{\Psi}_B(i\gamma^\mu D_\mu - m_B)\Psi_B,$$

$$D^\mu \equiv \partial^\mu - iA^\mu. \qquad (22.39)$$

To all orders in the loop expansion, the field A^μ and the gauge-fixing parameter α/e^2 remain free of renormalisations. The same holds for non-Abelian gauge theories. By absorbing the charge q (called coupling constant $g \equiv q$ henceforth) in the gauge field, the Lagrangian can be expressed as

$$\mathcal{L} = \frac{1}{2g_B^2}\text{Tr}(F_{\mu\nu}^2) + \frac{\alpha_B}{g_B^2}\text{Tr}(\partial_\mu A^\mu)^2 + \overline{\Psi}_B(i\gamma^\mu D_\mu - m_B)\Psi_B, \qquad (22.40)$$

where the gauge field and the gauge-fixing parameter receive no renormalisations; in other words they are already the renormalised field and gauge-fixing parameter. The field strength $F_{\mu\nu}$ and covariant derivative D_μ are now given by [compare this to Equations (18.35) and (18.41)]

$$D_\mu = \partial_\mu + A_\mu, \quad F_{\mu\nu} = \partial_\mu A_\nu - \partial_\nu A_\mu + [A_\mu, A_\nu]. \qquad (22.41)$$

If the gauge group is $SU(N)$ and there are n_f flavours of fermions the renormalisation of the coupling constant is given by (see Itzykson and Zuber Section 12-3-4)

$$g_B = \mu^{\frac{1}{2}\varepsilon} Z_g g = \mu^{\frac{1}{2}\varepsilon} \left(1 - \frac{(11N - 2n_f)g^2}{48\pi^2} \cdot \frac{1}{\varepsilon} + \cdots \right) g. \qquad (22.42)$$

As long as the number of fermion flavours is small enough, we see that the one-loop corrections to the bare coupling constant differ in sign from the equivalent expression for the Abelian case. It is the self-interactions of the non-Abelian gauge fields that are responsible for the asymptotic freedom of its running coupling constant. The running of the coupling is expressed in terms of the so-called beta function

$$\beta(g) \equiv \mu \frac{\partial g(g_B(\varepsilon), \mu, \varepsilon)}{\partial \mu}, \qquad (22.43)$$

where the derivative is taken at fixed ε and g_B ($g \equiv g_R$). For non-Abelian gauge theories one finds (μ_0 is an integration constant)

$$\beta(g) = -\frac{(11N - 2n_f)g^3}{48\pi^2} + \mathcal{O}(g^5),$$

$$\frac{1}{g^2(\mu)} = \frac{(11N - 2n_f)}{24\pi^2} \log(\mu/\mu_0) + \mathcal{O}(g^2(\mu)), \qquad (22.44)$$

whereas for QED (coupled to n_f flavours of fermions)

$$\beta(e) = \frac{n_f e^3}{12\pi^2} + \mathcal{O}(e^5), \quad \frac{1}{e^2(\mu)} = -\frac{n_f}{6\pi^2} \log(\mu/\mu_0) + \mathcal{O}(e^2(\mu)). \qquad (22.45)$$

For other regularisations the computation of the running coupling constant is similar, except that ε is replaced roughly by $1/\log(\Lambda)$. To the order displayed, the beta functions do not depend on the regularisation scheme.

It is perhaps appropriate to end these lecture notes with as classic an experimental test of renormalisation effects in field theory as the one for the Casimir energy in Chapter 2. It concerns the Lamb shift, measured in 1947, which is the very small energy splitting of the $2S_{\frac{1}{2}}$ and $2P_{\frac{1}{2}}$ orbitals in hydrogen atoms, receiving a contribution from vacuum polarisation effects (for a discussion of the other contributions, see Section 7-3-2 of Itzykson and Zuber). In Problem 39 it will be shown that to one-loop order the photon vacuum polarisation is given by [compare this to Equations (16.22) and (16.23); in the Landau gauge, $\alpha \to \infty$, we can drop the $\Lambda^{(\alpha)}$ factors]

$$\Sigma_{\mu\nu}(q) = -\Lambda^{(\alpha)}_{\mu\beta}(q)(q^2 g^{\beta\gamma} - q^\beta q^\gamma)\omega(q^2)\Lambda^{(\alpha)}_{\gamma\nu}(q). \qquad (22.46)$$

From the results of Problem 39, where ω is computed with Pauli–Villars regularisation, it can be deduced that (m is the electron mass)

$$\omega(q^2) = a \log(\Lambda/m) + b + cq^2 \quad \text{for} \quad q^2 \to 0, \ \Lambda \to \infty. \tag{22.47}$$

The precise values of the coefficients a and b are not very important, as the combination $a \log(\Lambda/m) + b$ can be absorbed in the field renormalisation (this means that a can be read off from Z_A given above). In Section 7-1-1 of Itzykson and Zuber it is shown that $c = e^2/(60\pi^2 m^2)$. In the static limit, as is relevant for the hydrogen atom, $q^2 = -\vec{q}^{\,2}$ and the photon exchange can be accurately described by the Coulomb potential

$$-\frac{e^2}{4\pi r} = -e^2 \int d_3\vec{q} \, \frac{e^{i\vec{q}\cdot\vec{r}}}{\vec{q}^{\,2}}, \tag{22.48}$$

which due to the vacuum polarisation is replaced by

$$-e^2 \int d_3\vec{q} \, \frac{e^{i\vec{q}\cdot\vec{r}}}{\vec{q}^{\,2}(1+\omega(-\vec{q}^{\,2}))} = -e^2 \int d_3\vec{q} \left(\frac{1}{\vec{q}^{\,2}} + c + \cdots \right) e^{i\vec{q}\cdot\vec{r}}$$

$$= -\frac{e^2}{4\pi r} - \frac{e^4}{60m^2\pi^2}\delta_3(\vec{r}). \tag{22.49}$$

The extra delta-function interaction, that arises from the vacuum fluctuations, only affects the wave functions that do not vanish in the origin. Consequently, only the energy of the S orbitals will be shifted by this correction

$$\Delta E(nS_{\frac{1}{2}}) = -\frac{4m\alpha_e^5}{15\pi n^3}, \quad \alpha_e = \frac{e^2}{4\pi}, \tag{22.50}$$

where n is the radial quantum number and α_e is the fine-structure constant.

23

Problems

DOI: 10.1201/b15364-23

1. **Violation of causality in $1+1$ dimensions**

   In the lecture notes it is shown that in $3+1$ dimensions the Hamiltonian $H = \sqrt{m^2c^4 + \vec{p}^2c^2}$, where $\vec{p} = -i\hbar\vec{\nabla}$, gives rise to violation of causality. In this exercise we will conclude that this is not a special property of this dimension by considering the $1+1$ dimensional case.

   (a) Give the exact plane wave solutions of Schrödinger's equation for the Hamiltonian $H = \sqrt{m^2c^4 + p^2c^4}$.

   (b) Let $\psi_0(x, t)$ be the solution of Schrödinger's equation with initial condition $\psi_0(x, 0) = \delta(x)$. It follows that $\psi(x, t) = \int dy f(y) \psi_0(x - y, t)$ is also a solution (for which initial condition?). Therefore it is sufficient to study the time evolution of ψ_0.

   Fourier expand $\psi_0(x, t)$ and rewrite this expression as

   $$\psi_0(x, t) = \frac{i}{c\pi} \partial_t K_0(z), \quad \text{with } K_0(z) = \int_0^\infty dy \cos\left(z \sinh(y)\right),$$

   $$z^2 = \frac{m^2c^2}{\hbar^2}(x^2 - c^2t^2).$$

   (K_0 is a modified Bessel function, whence the above expression can be rewritten in terms of ordinary Bessel functions. See, for example, Abromowitz and Stegun's handbook for details.)

   (c) Show that for $m \neq 0$ the solution violates causality.

   (d) Prove that $\mathrm{Re}(\psi_0) = (\psi_0 + \psi_0^*)/2$ respects causality but does not satisfy Schrödinger's equation. Show that ψ_0^* is a solution of the time-inverted Schrödinger equation, or equivalently Schrödinger's equation with opposite (negative) Hamiltonian. Prove that both ψ_0 and ψ_0^* satisfy the Klein–Gordon equation and that $\mathrm{Re}(\psi_0)$ is the unique solution that respects causality.

2. **Casimir effect**

   In quantum field theory the vacuum energy depends on the spatial volume. In the lecture notes it has been derived that a free scalar massless field which is spatially contained between two infinite parallel planes with separation x has an energy per unit area

$$E(x) = \frac{1}{2(2\pi\hbar c)^2} \sum_{m=1}^{\infty} \int d^2k \sqrt{k^2 + (\frac{\pi\hbar cm}{x})^2}.$$

This expression is divergent but can be made finite in a sensible way by subtraction of a corresponding slice in infinite volume, i.e., without boundary conditions in the x direction. An alternative way of getting rid of the unphysical infinite part is so-called dimensional regularisation. The above integral (the sum will be attacked analogously later) falls into a class of integrals that is parametrised by (a.o.) the dimension. The method then consists of computing the convergent integrals within this class, and redefining the divergent ones by analytic continuation (in the set of parameters) of the convergent outcomes.

For the case at hand, the following class of integrals is useful:

$$I_{n,\lambda,\mu}(\alpha) \equiv \int d^n k \frac{k^{2\lambda}}{(k^2 + \alpha^2)^\mu} \quad (n \in \mathbb{N}; \ \lambda, \mu \in \mathbb{C}; \ \alpha^2 > 0).$$

For $2\text{Re}(\lambda - \mu) + n < 0$ this is convergent (and analytic). For $n = 2$, $\lambda = 0$, $\mu = -1/2$ it reduces to the integral in $E(x)$. Assume for the time being that $2\text{Re}(\lambda - \mu) + n < 0$.

(a) Change to spherical coordinates and derive

$$I_{n,\lambda,\mu}(\alpha) = \pi B \left(1, \frac{1}{2}\right) B \left(\frac{3}{2}, \frac{1}{2}\right) \cdots B \left(\frac{n-1}{2}, \frac{1}{2}\right)$$

$$\times B \left(\lambda + \frac{n}{2}, \mu - \lambda - \frac{n}{2}\right) \alpha^{n+2\lambda-2\mu},$$

where B is the so-called beta function:

$$B(x, y) \equiv 2 \int_0^\infty dt \ t^{2x-1}(1 + t^2)^{-x-y}.$$

(b) Let the gamma function be defined as

$$\Gamma(x) = \int_0^\infty dt \ t^{x-1}e^{-t} \quad (\text{Re}(x) > 0).$$

Show that $\Gamma(x + 1) = x\Gamma(x)$ and $\Gamma(1/2) = \sqrt{\pi}$. Also prove

$$B(x, y) = \frac{\Gamma(x)\Gamma(y)}{\Gamma(x + y)}.$$

(c) Now write $I_{n,\lambda,\mu}(\alpha)$ in terms of gamma functions. Note that $\Gamma(x)$ can be continued analytically to $\text{Re}(x) \leq 0$ using $\Gamma(x+1) = x\Gamma(x)$. Therefore we can also continue I analytically to parameter values for which the original integral was divergent. Also notice that the dimension n can now be given arbitrary complex values without difficulties. Show that after having done these regularisations we obtain

$$E(x) = E_2(x), \quad E_n(x) = \sqrt{\pi}\hbar c \sum_{m=1}^{\infty} \left(\frac{\sqrt{\pi}m}{2x} \right)^{n+1} \frac{\Gamma\big(-(n+1)/2\big)}{\Gamma(-1/2)}.$$

(d) Only the summation over m remains to be regularised. Define the zeta function

$$\zeta(x) = \sum_{m=1}^{\infty} m^{-x} \quad (\mathrm{Re}(x) > 1).$$

For $x = -3$ this coincides with the relevant divergent summation. Like for the integrals, we would like to replace this expression by the analytic continuation of $\zeta(x)|_{\mathrm{Re}(x)>1}$ to $\mathrm{Re}(x) \le 1$. This continuation satisfies

$$\zeta(1-n) = \frac{(-1)^{n+1}B_n}{n}, \quad n \in \mathbb{N}, \tag{23.1}$$

where B_n are the Bernoulli numbers:

$$\sum_{n=0}^{\infty} B_n \frac{t^n}{n!} \equiv \frac{t}{e^t - 1}.$$

Derive eq. (23.1) by expanding $t/(e^t - 1)$ in e^{-t}, and e^{-t} in t (be careful with the t^0 term).

Hint: Introduce new parameters that enable change of summation order. In the end continue back to the relevant parameter values.

Finally compute the fully regularised energy per area $E(x)$ and pressure $F(x) = -dE(x)/dx$. Given the Bernoulli number $B_4 = -1/30$, evaluate this pressure for $x = 1\mu m$.

3. **Euler–Lagrange equation**

Let $\phi(x) = \phi(\vec{x}, t)$ be a complex scalar field with action functional $S = \int d^4x \mathcal{L}(x)$. $\mathcal{L}$ is the so-called Lagrange density [the Lagrangian is $L(t) = \int d^3\vec{x}\mathcal{L}(\vec{x}, t)$]:

$$\mathcal{L}(x) = \partial_\mu \phi^*(x)\partial^\mu \phi(x) - m^2 \phi^*(x)\phi(x),$$

with metric $g^{\mu\nu} = \mathrm{diag}(1, -1, -1, -1)$ and units such that $\hbar = 1$, $c = 1$.

(a) Prove by variational calculus the Euler–Lagrange equation (equation of motion) $\frac{\delta S}{\delta \phi(x)} = \partial_\mu \frac{\delta S}{\delta(\partial_\mu \phi(x))}$ and show that this gives the Klein–Gordon equation.

(b) Given the energy-momentum tensor

$$T_{\mu\nu}(x) = \partial_\mu\phi^*(x)\partial_\nu\phi(x) + \partial_\nu\phi^*(x)\partial_\mu\phi(x) - g_{\mu\nu}\mathcal{L}(x),$$

show that $\partial_\mu T^{\mu\nu} = 0$.

(c) Given the current density

$$J_\mu(x) = i\left(\phi^*(x)\partial_\mu\phi(x) - (\partial_\mu\phi^*(x))\phi(x)\right),$$

show that $\partial_\mu J^\mu(x) = 0$.

(d) Prove that the total energy E, momentum P_i and charge Q, given by $E(t) = \int d^3\vec{x} T_{00}(\vec{x}, t)$, $P_i(t) = \int d^3\vec{x} T_{0i}(\vec{x}, t)$ and $Q(t) = \int d^3\vec{x} J_0(\vec{x}, t)$, are conserved.

4. **Creation and annihilation operators**
   We start from operators p and q satisfying canonical commutation relations $[p, q] = -i\hbar$. Define

$$a = \frac{1}{\sqrt{2\hbar\omega}}(\omega q + ip), \quad a^\dagger = \frac{1}{\sqrt{2\hbar\omega}}(\omega q - ip), \quad N = a^\dagger a.$$

(a) Show that $[a, a^\dagger] = 1$. Also calculate the commutators $[a, N]$, $[a^\dagger, N]$ and $[(a^\dagger)^n, N]$.

(b) Define $|n\rangle$ by $N|n\rangle = n|n\rangle$, $\langle n|n\rangle = 1$. Show that

$$a|n\rangle = c_n^- |n-1\rangle, \qquad a^\dagger|n\rangle = c_n^+ |n+1\rangle, \qquad |n\rangle = c_n(a^\dagger)^n|0\rangle.$$

Compute c_n^-, c_n^+, c_n and show that they can be chosen real.

Given an algebra of operators and commutation relations, we mean by the associated Hilbert space the (smallest) Hilbert space that may be used to incorporate the algebra. What is the associated Hilbert space in the present case?

(c) Derive a matrix representation for the operators a, $a^\dagger$ and N.

(d) Now consider operators with <u>anti</u>commutation relations

$$\{b_r, b_s^\dagger\} = \delta_{r,s}, \quad \{b_r, b_s\} = \{b_r^\dagger, b_s^\dagger\} = 0,$$

where $\{X, Y\} \equiv XY + YX$. What is the corresponding Hilbert space?

Define $N_r = b_r^\dagger b_r$. What are the possible eigenvalues of N_r? Construct a matrix representation for the operators b_r, $b_r^\dagger$ and N_r. Why is the algebra generated by b_r and $b_r^\dagger$, with the above anticommutation relations, suitable for describing fermions?

Prove that exchanging b_r and $b_r^\dagger$ can be described by a unitary transformation.

(e) The BCS theory of superconductivity uses the following operators that describe annihilation and creation of electron pairs:

$$c_{\vec{k}} = b_{-\vec{k}\downarrow} b_{\vec{k}\uparrow}, \quad c_{\vec{k}}^{\dagger} = b_{\vec{k}\uparrow}^{\dagger} b_{-\vec{k}\downarrow}^{\dagger}.$$

Prove that $[c_{\vec{k}}, c_{\vec{k}'}] = [c_{\vec{k}}^{\dagger}, c_{\vec{k}'}^{\dagger}] = 0$ and calculate $[c_{\vec{k}}, c_{\vec{k}'}^{\dagger}]$. Also determine the Hilbert space and the action of the operators on this Hilbert space. Would you call the electron pairs fermions or rather bosons?

5. **Real and complex fields**
   Let us consider a real scalar field $\varphi(x)$ and a Hamiltonian

$$H = \int d^3\vec{x} \left\{ \frac{1}{2}\pi^2(x) + \frac{1}{2}\left(\partial_i\varphi(x)\right)^2 + \frac{1}{2}m^2\varphi^2(x) \right\},$$

where $\pi(x) = \partial_t\varphi(x)$ is the canonical momentum. For quantisation we postulate the following commutators at some time t, say $t = 0$. (Argue briefly why these relations are compatible with causality.)

$$[\pi(x), \pi(y)]\,|_{x_0=y_0=0} = [\varphi(x), \varphi(y)]\,|_{x_0=y_0=0} = 0;$$

$$[\pi(x), \varphi(y)]\,|_{x_0=y_0=0} = -i\delta_3(\vec{x} - \vec{y}).$$

Write the Fourier decomposition of $\varphi(x)$ as follows:

$$\varphi(x) = \frac{1}{(2\pi)^3} \int d^3\vec{k}\, \frac{1}{\sqrt{2k_0}} \left(a(\vec{k})e^{-ikx} + a^{\dagger}(\vec{k})e^{ikx} \right),$$

where $kx = k_0 t - \vec{k} \cdot \vec{x}$ and $k_0 = +\sqrt{\vec{k}^2 + m^2}$.

Remark: $\varphi(\vec{x}, t)$ is the Heisenberg representation of $\varphi(\vec{x}, 0)$. This can be verified explicitly at the end of part (d).

(a) Give the Fourier decomposition of $\pi(x)$. Why can we (formally) set $\pi(\vec{x}, x_0 = 0) = -i\partial/\partial\varphi(\vec{x}, x_0 = 0)$?

(b) Derive the commutation relations for $a(\vec{k})$ and $a^{\dagger}(\vec{k})$.

(c) What is the associated Hilbert space?

(d) Write the Hamiltonian H in terms of the occupation number (density) operators $N(\vec{k}) = a^{\dagger}(\vec{k})a(\vec{k})$. Note that H is time independent.

It is impossible to define a total charge Q for a real field $\varphi(x)$ (in a nontrivial way). Basically this is because a real field describes particles that are their own antiparticles. Therefore let us introduce a complex field $\varphi \neq \varphi^{\dagger}$ with Hamiltonian

$$H = \int d^3x \{\pi^{\dagger}(x)\pi(x) + \partial_i\varphi^{\dagger}(x)\partial_i\varphi(x) + m^2\varphi^{\dagger}(x)\varphi(x)\},$$

where $\pi(x) = \partial_t\varphi^{\dagger}(x)$, $\pi^{\dagger}(x) = \partial_t\varphi(x)$.

(e) Show that $H = \int d^3\vec{x}\, T_{00}$ (see Problem 3 for the definition of $T_{\mu\nu}$, in which classical fields now become operator fields).

The nontrivial commutators are postulated to be

$$[\pi(x), \varphi(y)]\,|_{x_0=y_0=0} = [\pi^\dagger(x), \varphi^\dagger(y)]\,|_{x_0=y_0=0} = -i\delta_3(\vec{x} - \vec{y}).$$

Let us write $\varphi(x) = [\varphi_1(x) + i\varphi_2(x)]/\sqrt{2}$ and substitute for the real fields $\varphi_i(x)$ the Fourier decompositions in terms of $a_i(\vec{k})$ and $a_i^\dagger(\vec{k})$.

(f) Give $a(\vec{k})$ and $b(\vec{k})$ in terms of $a_i(\vec{k})$ such that

$$\varphi(x) = \frac{1}{(2\pi)^3} \int d^3\vec{k} \frac{1}{\sqrt{2k_0}} \left(a(\vec{k})e^{-ikx} + b^\dagger(\vec{k})e^{ikx}\right).$$

Also derive the Fourier decompositions of $\varphi^\dagger(x)$, $\pi(x)$ and $\pi^\dagger(x)$.

(g) Give the mutual commutation relations for the operators $a(\vec{k})$, $a^\dagger(\vec{k})$, $b(\vec{k})$ and $b^\dagger(\vec{k})$.

(h) Write H in terms of $N^a(\vec{k}) = a^\dagger(\vec{k})a(\vec{k})$ and $N^b(\vec{k}) = b^\dagger(\vec{k})b(\vec{k})$.

We would like to interpret the particles created by $b^\dagger$ as the antiparticles of the ones created by $a^\dagger$. This allows us to define the total charge

$$Q = \text{const.}(\#\text{particles} - \#\text{antiparticles})$$
$$= \frac{e}{(2\pi)^3} \int d^3\vec{k} \left(N^a(\vec{k}) - N^b(\vec{k})\right).$$

(at $t = 0$).

(i) Prove that Q is conserved. Also show that Q can be written as

$$Q = \int d^3\vec{x}\rho(x) + \text{constant},$$

where $\rho(x) = -ie\{(\partial_t\varphi^\dagger)(x)\varphi(x) - \varphi^\dagger(x)(\partial_t\varphi)(x)\}$. Note that $\rho(x) = eJ_0(x)$ (see Problem 3).

6. **Commutation relations and causality**

We reconsider the Hermitian operator field

$$\varphi(x) = \frac{1}{(2\pi)^3} \int d^3\vec{k} \frac{1}{\sqrt{2k_0}} \left(a(\vec{k})e^{-ikx} + a^\dagger(\vec{k})e^{ikx}\right).$$

In Problem 5 commutation relations $\{[\varphi(x), \varphi(y)]\,|_{x_0=y_0=0} = 0 \text{ etc.}\}$ and a Hamiltonian have been introduced. Use these for deriving an integral representation for

$$\Delta(x - y) \equiv [\varphi(x), \varphi(y)];$$

x, y arbitrary.

Show that $\Delta(x-y) = 0$ whenever $x_0 = y_0$ (x_0 arbitrary). Also show that $\Delta(x - y)$ is Lorentz invariant and use this for generalising the result to x, y with $(x - y)^2 < 0$ (i.e., spatially separated).

Hint: Prove that $\int d^3\vec{k} = \int d^4k\delta(k^2 - m^2)\theta(k_0)2k_0$, where θ is the step function.

7. **Feynman rules for a classical field**

Consider real fields φ_1 and φ_2 as described by the Lagrangian

$$\mathcal{L}[\varphi_1, \partial_\mu\varphi_1, \varphi_2, \partial_\mu\varphi_2] = \frac{1}{2}\partial_\mu\varphi_1\partial^\mu\varphi_1 + \frac{1}{2}\partial_\mu\varphi_2\partial^\mu\varphi_2$$
$$- g_0 \log\left[1 + \frac{1}{2}g_1(\varphi_1 - F)^2\right] - \frac{1}{2}g_2\varphi_1\varphi_2^2.$$

(a) Determine the dimension of the fields φ_i and the constants g_i, F. (Remember that for $\hbar = 1$, $c = 1$ all dimensions are powers of $[l] = [m]^{-1}$; also the action $S = \int d^4x\mathcal{L}(x)$ is dimensionless.)

(b) In a perturbative calculation $\tilde{\varphi} = \varphi_1 - F$ and φ_2 are chosen as fundamental fields. Explain why.

Expand $\mathcal{L}$ in $\tilde{\varphi}$ and φ_2 (up to 4th-order terms). Write the result as $\mathcal{L} = \mathcal{L}_0 + \mathcal{L}_{\text{int}}$, where $\mathcal{L}_0$ are the quadratic terms and $\mathcal{L}_{\text{int}}$ contains the interaction terms. What are the masses of the fields $\tilde{\varphi}$ and φ_2?

(c) We now introduce source terms $-\tilde{J} \cdot \tilde{\varphi}$ and $-J_2 \cdot \varphi_2$. Derive the Feynman rules for the perturbative expansion using the (classical) method in the lecture notes (pp. 19–22). Use the following notation:

$$\underline{\qquad\qquad}\ \tilde{G} \qquad\qquad \underline{\qquad\qquad}\times\ \tilde{J}$$

$$------\ G_2 \qquad\qquad -----\circ\ J_2 \qquad \text{etc.}$$

(d) Which expression is associated to the diagram below?

8. **Photon propagator**

Gauge invariance complicates the derivation of the photon propagator (see lecture notes p.22–23). In this exercise we will fix the gauge by using the following Lagrangian:

$$\mathcal{L}(x) = -\frac{1}{4}F_{\mu\nu}(x)F^{\mu\nu}(x) - \lambda(x)\partial_\mu A^\mu(x) - J_\mu(x)A^\mu(x).$$

This describes a photon field A_μ and a Lagrange multiplier λ in the presence of an external (i.e., not dynamical) source J_μ.

(a) Use partial integration to write the quadratic part of the action as

$$\frac{1}{2}((A_\mu), \lambda) \cdot \hat{M}\begin{pmatrix}(A_\nu)\\ \lambda\end{pmatrix},$$

where $\hat{M}$ is a <u>Hermitian</u> 5×5 matrix operator. The inner product '·' includes an integration over space-time.

(b) Show that $\hat{M}$ is invertible and that the corresponding photon propagator is the same as in the so-called Landau gauge $\alpha \to \infty$ (lecture notes p. 23).

<u>Hint</u>: Work in Fourier space.

9. **Coulomb gauge and temporal gauge**
The gauge freedom of the photon field can be eliminated through an extra constraint besides the equations of motion (imposing the constraint is usually called 'choosing a gauge'). Examples:

(1) Lorentz gauge $\partial_\mu A^\mu = 0$
(2) Coulomb gauge $\partial_i A_i \equiv \vec{\nabla} \cdot \vec{A} = 0$
(3) temporal gauge $A_0 = 0$.

Here we will analyze the conditions (2) and (3). These are not Lorentz invariant, but expose the photon's degrees of freedom nicely.

(a) Show that (2) or (3) can always be realised after an appropriate gauge transformation $A_\mu(x) \to A_\mu(x) + \partial_\mu \Lambda(x)$. Furthermore show that (2) and (3) can be imposed simultaneously in the absence of sources (i.e., $J_\mu = 0$).

(b) The transversal (T) and longitudinal (L) components of an arbitrary vector field $\vec{v}$ are defined as follows:

$$\vec{v} = \vec{v}_T + \vec{v}_L, \quad \vec{\nabla}\cdot\vec{v}_T = 0, \quad \vec{\nabla}\times\vec{v}_L = \vec{0}$$

Write down the relations between $\vec{A}$ and $\vec{E}$, $\vec{B}$ in terms of their T and L components. Also express Maxwell's equations in these components. Here $\vec{A}$, $\vec{E}$ and $\vec{B}$ stand for the vector potential, electric field and magnetic field, respectively.

(c) <u>Coulomb gauge</u>
Show that $A_0(\vec{x}, t)$ is completely determined by $\rho(\vec{x}', t)$ and the spatial boundary conditions at time t (hence the name

'instantaneous Coulomb potential'). It follows that the longitudinal component of the physical field $\vec{E}$ is not a degree of freedom in the radiation field. (Why? What do we mean exactly by a degree of freedom in a classical system?)

<u>Temporal gauge</u>

Show that $\vec{A}_L$ is completely determined by the charge distribution ρ and the spatial boundary conditions, together with an initial condition $\vec{A}_L(\vec{x}, t \to -\infty)$. Show (again) that $\vec{E}_L$ is not a degree of freedom in the radiation field.

10. **Preparation for the path integral**

Consider a one-dimensional harmonic oscillator with the Hamiltonian $H = \frac{\hat{p}^2}{2m} + \frac{1}{2}m\omega^2\hat{q}^2$. Here $\hat{p} = \frac{1}{i}\frac{\partial}{\partial q}$ and $\hat{q}$ is the position operator, so that $< q|H|p > = \frac{1}{\sqrt{2\pi}}e^{ipq}h(p, q)$ with $h(p, q) = \frac{p^2}{2m} + \frac{1}{2}m\omega^2q^2$.

(a) Prove the following **exact** identity ($\delta t \equiv T/n$):

$$K_n(q_n, q_0, T) \equiv \int \frac{dp_n}{2\pi} \prod_{i=1}^{n-1} \frac{dq_i dp_i}{2\pi}$$

$$\times \exp\left[i\sum_{j=1}^{n}\{p_j(q_j - q_{j-1}) - h(p_j, q_j)\delta t\}\right]$$

$$= < q_n|\left\{\exp\left(-i\frac{m\omega^2}{2}\hat{q}^2\delta t\right)\exp\left(-i\frac{\hat{p}^2\delta t}{2m}\right)\right\}^n|q_0 > .$$

(b) Show that $K_n(q_n, q_0, T) =< q_n|\exp(-i\frac{m\omega^2}{4}\hat{q}^2\delta t)\mathcal{T}^n\exp(i\frac{m\omega^2}{4}\hat{q}^2\delta t)|q_0 >$, where $\mathcal{T} = \exp(-i\frac{m\omega^2}{4}\hat{q}^2\delta t)\exp(-i\frac{\hat{p}^2\delta t}{2m})\exp(-i\frac{m\omega^2}{4}\hat{q}^2\delta t)$. Prove that $\mathcal{T}$ is a unitary operator.

(c) We are going to prove that $\mathcal{T} = \exp(-i\tilde{H}\delta t)$, where $\tilde{H}$ is also a harmonic oscillator Hamiltonian: $\tilde{H} = \frac{\hat{p}^2}{2M} + \frac{1}{2}M\Omega^2\hat{q}^2$.

<u>NOTE</u>: Until that is proven, one should of course use $\mathcal{T}$ as defined in part (b).

1. Show that $[\hat{p}, \hat{q}] = -i$ implies

$$[e^{\alpha\hat{p}^2}, \hat{q}] = -2i\alpha\hat{p}e^{\alpha\hat{p}^2} \text{ and } [e^{\alpha\hat{q}^2}, \hat{p}] = 2i\alpha\hat{q}e^{\alpha\hat{q}^2}$$

for any $\alpha \in \mathbb{C}$. Now solve the 'eigenvalue equation'

$$\mathcal{T}(\kappa_{\pm}\hat{q} + \lambda_{\pm}\hat{p}) = \mu_{\pm}(\kappa_{\pm}\hat{q} + \lambda_{\pm}\hat{p})\mathcal{T}$$

($\kappa_{\pm}, \lambda_{\pm}, \mu_{\pm} \in \mathbb{C}$). Show that $\mu_{\pm} = \exp(\pm i\Omega\delta t)$, with Ω defined by $\sin(\frac{1}{2}\Omega\delta t) = \frac{1}{2}\omega\delta t$.

2. Determine the commutation relations between $\kappa_{+}\hat{q} + \lambda_{+}\hat{p}$ and $\kappa_{-}\hat{q} + \lambda_{-}\hat{p}$. For which normalisation are we dealing with creation

and annihilation operators $\hat{a}^\dagger$, $\hat{a}$? Show that the corresponding Hamiltonian is given by $\tilde{H}$, with $M = m\sin(\Omega\delta t)/(\Omega\delta t)$.

3. Now that $(\kappa, \lambda, \mu)_\pm$ are known, the eigenvalue equation determines $\mathcal{T}$ uniquely up to a $\hat{p}, \hat{q}$ independent factor. Prove that $\mathcal{T} = C\exp(-i\hat{a}^\dagger\hat{a}\Omega\delta t)$ satisfies the equation ($C \in \mathbb{C}$). Use the definition of $\mathcal{T}$ to show that $C = \exp(-\frac{1}{2}i\Omega\delta t)$.

   Hint: Since C is independent of $\hat{p}, \hat{q}$, it can be determined by calculating $\langle 0|\mathcal{T}|0\rangle$ with $|0\rangle$ the $\tilde{H}$-vacuum ($\hat{a}|0\rangle \equiv 0$). First evaluate $\langle p|\exp(-i\frac{m\omega^2}{4}\hat{q}^2\delta t)|0\rangle = A\exp(-Bp^2/2)$, with A and B defined appropriately.

(d) Use the above result to show that $\lim_{n\to\infty} K_n(q_n, q_0, T) = < q_n|e^{-iHT}|q_0 >$.

11. **Path integral for a free particle**

    We start from the path integral for the evolution operator associated to Schrödinger's equation (lecture notes p. 30). As Lagrangian we take $L(q, \dot{q}) = \frac{1}{2}m\dot{q}^2$, and problems from integrating rapidly oscillating functions are avoided by choosing so-called Euclidean time $\tau = iT$. The path integral then becomes (with $d\tau \equiv \tau/n$ and for $n \to \infty$):

$$\langle q'|U(\tau)|q\rangle = \left[\frac{m}{2\pi\delta\tau}\right]^{n/2} \left(\prod_{j=1}^{n-1}\int dq_j\right) e^{-S(q_0, q_1, \ldots, q_n)}, \qquad (23.2)$$

where $q_0 \equiv q$ and $q_n \equiv q'$, and with action

$$S(q_0, q_1, \ldots, q_n) = \sum_{j=1}^{n}\frac{m}{2}\left[\frac{q_j - q_{j-1}}{\delta\tau}\right]^2\delta\tau.$$

The Euclidean evolution operator is $U(\tau) = \exp(-H\tau)$, H being the usual quantum mechanical Hamiltonian associated to L.

Upon defining

$$U(\tilde{q}, \tau) \equiv \langle\tilde{q}|U(\tau)|0\rangle \overset{\text{transl. inv.}}{=} \langle q'|U(\tau)|q\rangle, \quad \tilde{q} \equiv q' - q,$$

$U(\tilde{q}, \tau)$ satisfies the Euclidean Schrödinger equation by construction. Due to the Euclidean time this is a diffusion equation:

$$U(\tilde{q}, 0) = \delta(\tilde{q}), \quad \frac{\partial}{\partial\tau}U(\tilde{q}, \tau) = \frac{1}{2m}\frac{\partial^2}{\partial\tilde{q}^2}U(\tilde{q}, \tau) \qquad (23.3)$$

(a) Determine $U(\tilde{q}, \tau)$ by solving eq. (23.3) (use a Fourier transform).

(b) In this simple case the path integral in eq. (23.2) can be calculated explicitly. We will do this by changing variables

$$y_j = q_j - q_{j-1}, \quad j = 1, 2, \cdots, n.$$

Show that $\prod_{j=1}^{n-1}dq_j = (\prod_{j=1}^{n}dy_j)\delta(\tilde{q} - \sum_{j'=1}^{n}y_{j'})$.

The δ function can be written as

$$\delta\left(\tilde{q} - \sum_{j=1}^{n} y_j\right) = \frac{1}{2\pi} \int d\omega \exp\left(i\omega\left(\sum_{j=1}^{n} y_j - \tilde{q}\right)\right).$$

These steps reduce the path integral to a product of Gaussian integrals. Perform the integrations and verify that the outcome equals the result in (a).

12. **Massive vector fields**

The following Lagrangian (mass $m \neq 0$) describes a massive vector field,

$$\mathcal{L} = -\tfrac{1}{4}F_{\mu\nu}F^{\mu\nu} + \tfrac{1}{2}m^2 A_\mu A^\mu.$$

(a) Show that this Lagrangian is <u>not</u> gauge invariant.

(b) Determine the equations of motion for the field $A_\mu(x)$. Show that these are equivalent to

$$\partial_\mu A^\mu = 0 \quad (*) \quad , \quad (\partial^2 + m^2)A_\mu = 0.$$

<u>Remark</u>: The condition $(*)$, being a gauge choice in the massless case (see Problem 8), is now imposed by the equations of motion!

(c) Bring $\mathcal{L}$ to the form $\tfrac{1}{2}A_\mu M^{\mu\nu} A_\nu$ (more precisely, use partial integration to find an M such that this gives the same action—and therefore the same equations of motion). Construct the inverse of the operator M (use a Fourier transform).

(d) Now add a source term: $-J_\mu A^\mu$ with $\partial_\mu J^\mu = 0$. Which expression for $A_\mu(k)$ is associated to the following Feynman diagram?

$$- - - - - - - \times$$

Are there other diagrams in this model contributing to $A_\mu(k)$?

$(M^{-1})_{\mu\nu}$ consists of two terms. Show that one of them drops out of $(M^{-1})_{\mu\nu}(k)J^\nu(k)$, and that $\lim_{m\to 0}(M^{-1})_{\mu\nu}(k)J^\nu(k)$ exists. Compare this limit to the Maxwell propagator (lecture notes p. 23) for $\alpha = 1$, the so-called Feynman gauge.

13. **Perturbative approach to the path integral**

In this exercise we will treat perturbatively the generating function $Z(J)$ for a real scalar φ^3 theory. The Lagrangian reads

$$\mathcal{L} = \mathcal{L}_2 - V_{\text{int}} - J\varphi, \quad \text{with } V_{\text{int}} = \frac{g}{3!}\varphi^3, \quad \mathcal{L}_2 = \tfrac{1}{2}\varphi G^{-1}\varphi,$$

$$G^{-1} = -(\partial_\mu \partial^\mu + m^2).$$

In the lecture notes (p. 52) it has been shown that the path integral can be reduced to

$$Z(J) = e^{-i \int d^4x\, V_{\text{int}}(i\frac{\delta}{\delta J(x)})} e^{-\frac{i}{2} \int d^4y \int d^4z\, J(y) G(y,z) J(z)}.$$

(a) Show that

$$Z(J) = 1 + \left(-\frac{i}{2}\ \times\!\!-\!\!\!\times\ -\ \frac{1}{8}\ \begin{array}{c}\times\!\!-\!\!\times\\ \times\!\!-\!\!\times\end{array} \right)$$

$$+ \left(\frac{1}{2}\ \bigcirc\!\!-\!\!\times\ -\ \frac{i}{3!}\ \begin{array}{c}\times\quad\times\\ \curlyvee\\ \times\end{array}\ -\ \frac{i}{4}\ \bigcirc\!\!-\!\!\times \right)$$

$$+ \mathcal{O}(J^5) + \mathcal{O}(g^2).$$

Here $\times\!\!-\!\!\times = \int d^4x \int d^4y\, J(x) G(x,y) J(y)$ etc. (do not work in Fourier space).

N.B. In this exercise you are not supposed to work out the analytical expressions associated to the Feynman diagrams.

(b) Read pages 103–105 from *Diagrammar* (cds.cern.ch/record/186259/files/CERN-73-09.pdf) carefully. Verify that the combinatorial factors in the above expression are correctly given by the *Diagrammar* prescription.

Remark: In this prescription the sources J should be considered as 1-vertices.

(c) Show up to first order in g and fourth order in J that $Z(J) = \exp[G(J)]$, $G(J)$ being the sum of <u>connected</u> diagrams.

(d) The 'n-point function' can be expressed in the following way:

$$\langle \varphi_n \cdots \varphi_1 \rangle \equiv \langle 0|\varphi_n \cdots \varphi_1|0 \rangle = \frac{1}{Z(0)} \left(i\frac{\delta}{\delta J_1} \cdots i\frac{\delta}{\delta J_n} Z(J) \right)_{J=0}$$

[$\varphi_i \equiv \varphi(\vec{x}_i, t_i)$, $t_{i+1} > t_i$, $|0\rangle =$ ground state in absence of J]. This is why $Z(J)$ is called the generating function.
Substitute the result of part (a) to obtain the one-, two-, three- and four-point functions up to order g. Argue that the product rule guarantees that the *Diagrammar* prescriptions for diagrams in $Z(J)$ and $\langle \varphi_n \cdots \varphi_1 \rangle$ are consistent, and verify explicitly the correctness of the Diagrammar prescription for the one-, two-, three- and four-point functions to the given order in g.

(e) Show nonperturbatively that

$$\left(i\frac{\delta}{\delta J_1} i\frac{\delta}{\delta J_2} G(J) \right)_{J=0} = \langle \varphi_2 \varphi_1 \rangle - \langle \varphi_2 \rangle \langle \varphi_1 \rangle.$$

For $n > 2$ similar expressions hold. This means that $G(J)$ is the generator of quantum fluctuations.

14. **Combinatorial factors**

    (a) Given a real scalar field φ with interaction

$$V_{\text{int}} = \frac{\alpha}{3!}\varphi^3 + \frac{\beta}{4!}\varphi^4,$$

determine the combinatorial factors of the following diagrams [see the discussion on pp. 54 to 56 or the section on combinatorial factors in 'Diagrammar', CERN Yellow report 73-9, by G. 't Hooft and M. Veltman, reprinted in *Under the Spell of the Gauge Principle* by G. 't Hooft (World Scientific, Singapore, 1994)]

$$\underline{1}\text{---}\underline{2} \; ; \; \underline{1}\text{---}\underline{2} \; ; \; ^2\!\!\!\bigvee^3_{|1} \; ; \; \underline{1}\text{--O--O--}\underline{2} \; ; \; \ominus \; ; \; \infty$$

    (b) Consider the following models:

    I Scalar field A,

$$\mathcal{L} = \tfrac{1}{2}\partial_\mu A\partial^\mu A - \tfrac{1}{2}m^2 A^2 - \frac{\lambda}{3!}A^3 - JA.$$

II Scalar fields A and B with equal mass,

$$\mathcal{L} = \tfrac{1}{2}\partial_\mu A\partial^\mu A + \tfrac{1}{2}\partial_\mu B\partial^\mu B - \tfrac{1}{2}m^2 A^2 - \tfrac{1}{2}m^2 B^2 - \frac{\mu}{2!}A^2 B - JA.$$

We limit ourselves to diagrams with an even number of external A lines (and no external B's). Let us pose the question whether we can make a distinction between the above models from knowledge of the amplitudes for its diagrams.

1. Let us first consider tree diagrams.
    - Show that λ and μ can be chosen such that the models I and II give an identical four-point function:

    - Show that the six-point function is different for the models I, II.

2. Proceed to show that at one-loop level even the two-point function, which at tree level is trivially the same, is different for the models I, II.

15. **Quantum corrections**

    In this exercise we set out to prove that the expansion of Feynman diagrams in the number of loops amounts to an expansion in powers of $\hbar$. We consider a real scalar field theory with an arbitrary interaction potential:

$$V_{\text{int}} = \sum_{n \geq 3} \frac{g_n}{n!} \varphi^n.$$

To each Feynman diagram we associate the following quantities: E, I, L and V_n (number of external lines, internal lines, loops, and vertices with n lines, respectively).

(a) Prove that for any connected diagram the following relations hold:

$$\begin{cases} L = I + 1 - \sum_{n \geq 3} V_n \\ \sum_{n \geq 3} n V_n = E + 2I. \end{cases}$$

Hint: Any diagram can be reduced to a tree diagram (i.e., a diagram with no loops) by cutting L times appropriate internal lines (this is the precise definition of L). Determine how E, I, L, $\sum V_n$ and $\sum n V_n$ have changed after one such cut. Another operation is the amputation of an external leg. Find the change in the above quantities for this operation too. Finally determine these quantities for a simple diagram in order to obtain the 'initial condition'.

(b) Since we are looking for quantum effects, we do <u>not</u> take $\hbar = 1$ (for convenience we keep $c = 1$, though). Powers of $\hbar$ can now pop up at several places in the Lagrangian. We can limit the number of such places by conveniently choosing the dimensions of φ, J and $\{g_n\}$. Show that this can be done in such a way that

$$\mathcal{L}_J \overset{\hbar=1}{\equiv} \mathcal{L} - J\varphi \overset{\hbar \neq 1}{=} \tfrac{1}{2} \partial_\mu \varphi \partial^\mu \varphi - \tfrac{1}{2} \frac{m^2}{\hbar^2} \varphi^2 - \sum_{n \geq 3} \frac{g_n}{n!} \varphi^n - J\varphi,$$

but that the $\hbar$ dependence in the quadratic part cannot be removed.

NOTE: For $\hbar \neq 1$, mass and 1/length have independent dimensions.

Remark: It is natural to require that the classical theory (i.e., the Euler–Lagrange equations) is independent of $\hbar$. The above result then implies that $m \sim \hbar$, so that even in the classical theory the mass is an effective parameter of quantum mechanical origin.

Show that the path integral now reads

$$Z(J) = C \int \mathcal{D}\varphi e^{\frac{i}{\hbar} \int d^4x \left[\mathcal{L}(\varphi(x)) - J(x)\varphi(x) \right]}$$

(C independent of J).

(c) We absorb the factor $\hbar$ in exponent of $Z(J)$ into the quadratic part of $\mathcal{L}$ by defining

$$\tilde{\varphi} = \hbar^{-1/2}\varphi, \quad \tilde{J} = \hbar^{-1/2}J.$$

This gives

$$Z(J) = \tilde{C} \int \mathcal{D}\tilde{\varphi} e^{i \int d^4x \left[\tilde{\mathcal{L}}(\tilde{\varphi}(x)) - \tilde{J}(x)\tilde{\varphi}(x) \right]}.$$

What is the expression for $\tilde{\mathcal{L}}(\tilde{\varphi})$? Show that the propagator for the field $\tilde{\varphi}$ does not have any $\hbar$ dependence. This means that all factors of $\hbar$ in a diagram come from the vertices (and external lines). Express the total power of $\hbar$ in terms of $\{V_n\}$ (and E). Finally make use of the results in (a) to prove that this power equals L, up to a function of E alone.

(d) Show that for a model with only four-point interactions ($g_n = 0$ for $n \neq 4$) the expansion in the number of loops L can be interpreted as an expansion in powers of g_4.

16. **Legendre transformation and classical limit**
In this exercise we will consider the connection between quantum field theory and classical field theory once more. Therefore take $\hbar \neq 1$ again. As explained in the previous exercise, we then have

$$Z[J] = \int \mathcal{D}\varphi e^{\frac{i}{\hbar} \int d^4x (\mathcal{L} - J\varphi)} = \int \mathcal{D}\varphi e^{\frac{i}{\hbar} S[J, \varphi]}, \qquad (23.4)$$

where $S[J, \varphi]$ is independent of $\hbar$. Furthermore, $Z[J] = \exp(G[J]/\hbar)$, $\frac{G[J]}{\hbar}$ being the sum of <u>connected</u> diagrams. The overall factor of $\hbar$ has been conveniently chosen $1/\hbar$ so that

$$G[J] = \sum_{L=0}^{\infty} \hbar^L G_{(L)}[J], \quad (L = \#\text{loops})$$

with $\hbar$-independent $G_{(L)}[J]$'s. A saddle point or stationary phase approximation ($\hbar \to 0$) of eq. (23.4) then immediately gives

$$G_{(0)}[J] = iS[J, \varphi_{\text{cl}}[J]]. \qquad (23.5)$$

Here $\varphi_{\text{cl}}[J]$ is the solution of the Euler–Lagrange equations $\delta S[J, \varphi]/\delta\varphi = 0$ (hence the subscript 'cl', for 'classical'). This saddle point $\varphi_{\text{cl}}[J]$ is unique under the assumption that in eq. (23.4) only fields vanishing (sufficiently fast) at infinity are integrated over.

Let us inspect if eq. (23.5) is reproduced by perturbation theory. For convenience we limit ourselves to φ^3 theory, whose Lagrangian has already been introduced in Problem 13.

(a) 1. Substitute the expansion of $\varphi_{cl}[J]$, as given on p. 21 of the lecture notes, in the action in order to obtain

$$-S\big[J, \varphi_{cl}[J]\big] = \tfrac{1}{2}\;\times\!\!-\!\!\times\; + \tfrac{1}{6}\;\\; + \tfrac{1}{8}\;\\; + \mathcal{O}(J^5).$$

Verify explicitly that eq. (23.5) holds to this order in J.

2. It follows from the path integral that [compare this to exercise 13(d)]

$$\langle\varphi\rangle[J](x) = i\,\frac{\delta G[J]}{\delta J(x)}. \tag{23.6}$$

Here $\langle\varphi\rangle[J]$ stands for the expectation value of the Heisenberg operator $\hat{\varphi}(x) = \hat{\varphi}(\vec{x}, t)$ in the groundstate $|0\rangle[J]$ of the Hamiltonian $\hat{H}[J]$, i.e., in the presence of a source J. Note that $\langle\varphi\rangle[J]$ is an ordinary real valued field, and not an operator field. Also note that in each point x it is a functional of J.

Show up to third order in J that

$$\langle\varphi\rangle[J] = \varphi_{cl}[J] + \mathcal{O}(\hbar). \tag{23.7}$$

(b) Now let us see if we can generalise these results to arbitrary order in J. Brute force as used in (a) is of no use here because this method generates the combinatorial factors for $S[J, \varphi_{cl}[J]]$ in an almost intractable way. The proper framework for the proof is the formalism of Legendre transformations (see Itzykson and Zuber for more details).

We assume that eq. (23.6) is invertible to $J(x) = J[\langle\varphi\rangle](x)$. This allows us to define a functional Γ on $\langle\varphi\rangle$ via a Legendre transform:

$$i\Gamma[\langle\varphi\rangle] \equiv G[J[\langle\varphi\rangle]] + i(J[\langle\varphi\rangle], \langle\varphi\rangle), \tag{23.8}$$

with $(f, g) \equiv \int d^4x f(x)g(x)$. Derive from eq. (23.6) that

$$\frac{\delta\Gamma[\langle\varphi\rangle]}{\delta\langle\varphi\rangle(x)} = J[\langle\varphi\rangle](x). \tag{23.9}$$

Hint: The chain rule for functional derivatives reads:

$$\frac{\delta f[g[h]]}{\delta h(x)} = \int d^4y\,\frac{\delta f[g]}{\delta g(y)}\bigg|_{g=g[h]}\frac{\delta g[h](y)}{\delta h(x)}.$$

Remark: An important example of a Legendre transform is the relation between a Lagrangian and its Hamiltonian: $H(q, p) = p\dot{q}(p) - L(q, \dot{q}(p))$ with $p = \partial L(q, \dot{q})/\partial \dot{q}$. (The position q plays no role in this transformation.)

(c) 1. It is useful to Taylor expand $G[J]$ around $J = 0$:

$$G[J] = \sum_{n=1}^{\infty} \frac{(-i)^n}{n!} \int d^4x_1 \cdots d^4x_n G^{(n)}(x_1, \cdots, x_n) J(x_1) \cdots J(x_n).$$

Why can we disregard the $n = 0$ contribution?

NOTE: $G^{(n)}$ is precisely the connected n-point function as defined in Problem 13.
We also expand $\Gamma[\langle\varphi\rangle]$ around $\langle\varphi\rangle[J = 0]$. For simplicity we limit ourselves to the case $\langle\varphi\rangle[0] = 0$.

$$\Gamma[\langle\varphi\rangle] = \sum_{n=2}^{\infty} \frac{1}{n!} \int d^4x_1 \cdots d^4x_n \Gamma^{(n)}(x_1, \ldots, x_n)\langle\varphi\rangle(x_1) \cdots \langle\varphi\rangle(x_n).$$

Why do the $n = 0, 1$ terms vanish?

2. $\Gamma^{(n)}$ can be obtained from $\{G^{(m)|m\le n}\}$ by differentiating eq. (23.6) $n - 1$ times with respect to $\langle\varphi\rangle$ and then setting $\langle\varphi\rangle = 0$ (corresponding to $J[\langle\varphi\rangle] = 0$). Show that

$$\Gamma^{(2)} = i(G^{(2)})^{-1}, \quad \Gamma^{(3)} = -iG^{(3)\,\mathrm{amp}},$$

where 'amp' means amputation:

$$G^{(n)\mathrm{amp}}(x_1, \cdots, x_n) \equiv \int \prod_{i=1}^{n} \left(d^4y_i(G^{(2)})^{-1}(x_i, y_i)\right) G^{(n)}(y_1, \ldots, y_n).$$

Argue that $G^{(3)\mathrm{amp}} = G^{(3)1\mathrm{PI}}$. The latter stands for the sum of '1 particle irreducible' diagrams, i.e., amputated diagrams that are still connected after cutting one arbitrary internal line. In general the following holds:

$$\Gamma^{(n)} = -iG^{(n)1\mathrm{PI}}. \quad (n \ge 3)$$

You are not asked to prove this, but it might be enlightening to check it for $n = 4$.

3. Use the above to show that, to order $\hbar^0$,

$$\Gamma^{(2)}_{(0)}(x, y) = -\delta_4(x-y)(\partial_\mu\partial^\mu + m^2), \quad \Gamma^{(3)}_{(0)}(x_1, x_2, x_3)$$
$$= -g\delta_4(x_1-x_2)\delta_4(x_2-x_3),$$

whereas $\Gamma^{(n)}_{(0)} = 0$ for $n \ge 4$. Also show that

$$\Gamma_{(0)}[\langle\varphi\rangle] = S[J = 0, \langle\varphi\rangle].$$

(d) Show that, to 0th order in $\hbar$, eq. (23.9) is just the Euler–Lagrange equation (for $\langle\varphi\rangle$). Under what boundary conditions can you now prove eq. (23.7)? Finally prove eq. (23.5).

Remark: The above shows that $\Gamma[\langle\varphi\rangle]$ may be viewed as a quantum mechanical generalisation of the classical action (without source term). The physical relevance of this particular generalisation comes from eq. (23.9). Apparently the observable $\langle\varphi\rangle[J]$ is governed by this generalised Euler–Lagrange equation. The quantum corrections usually cause $\langle\varphi\rangle[J] \neq \varphi_{cl}[J]$. For $J = 0$ a symmetry often prohibits such a shift, but for $J \to 0$ the shift may still be possible. In such a case $\langle\varphi\rangle[J \to 0]$, and therefore $|0\rangle[J \to 0]$, is less symmetric than $\varphi_{cl}[J \to 0]$. This means that quantum fluctuations can (spontaneously) break a symmetry.

17. **Feynman rules for complex fields**

If two real scalar fields, φ_1 and φ_2, are governed by the Lagrangian

$$\mathcal{L}(\varphi_1, \varphi_2) = \tfrac{1}{2}\partial_\mu\varphi_1\partial^\mu\varphi_1 + \tfrac{1}{2}\partial_\mu\varphi_2\partial^\mu\varphi_2 - \tfrac{1}{2}m^2(\varphi_1^2 + \varphi_2^2)$$
$$-V(\varphi_1^2 + \varphi_2^2) - J_1\varphi_1 - J_2\varphi_2$$

then it is possible to give an equivalent formulation using the complex fields

$$\varphi = \tfrac{\varphi_1+i\varphi_2}{\sqrt{2}}; \quad \varphi^* = \tfrac{\varphi_1-i\varphi_2}{\sqrt{2}}; \quad J = \tfrac{J_1+iJ_2}{\sqrt{2}}; \quad J^* = \tfrac{J_1-iJ_2}{\sqrt{2}}$$

which transforms to the Lagrangian

$$\mathcal{L}(\varphi, \varphi^*) = \partial_\mu\varphi\partial^\mu\varphi^* - m^2\varphi^*\varphi - V(2\varphi^*\varphi) - J^*\varphi - J\varphi^*$$

(see Problem 5 for the interpretation of φ and φ^* in terms of particles and antiparticles). Among the Feynman rules we now find oriented lines:

(a) Which <u>two</u> processes are described by this diagram?

(b) Give all Feynman rules for the model with

$$V(2\varphi^*\varphi) = \tfrac{1}{4}g(\varphi^*\varphi)^2.$$

(c) For this potential write down all connected diagrams with at most two loops contributing to

 .

18. **Elementary scalar processes**

Consider three real scalar fields (A, B, C) described by the Lagrangian

$$\mathcal{L} = \tfrac{1}{2}\left(\partial_\mu A \partial^\mu A + \partial_\mu B \partial^\mu B + \partial_\mu C \partial^\mu C - m_A^2 A^2 - m_B^2 B^2\right.$$
$$\left. - m_C^2 C^2 - g_A A^2 C - g_B B^2 C\right).$$

(a) If $m_C > 2m_A$ a C particle can decay into two A particles. To lowest order (in the couplings) this process is associated to the Feynman diagram

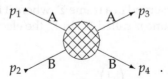

Determine the S-matrix element $_{\text{out}}\langle p_1 p_2 | q\rangle_{\text{in}}$ to this order (lecture notes p. 61). Also give an expression for the decay width $\Gamma(C \to 2A)$. Work out this expression for a C particle at rest ($\vec{q} = \vec{0}$). Determine the behaviour of $\Gamma(C \to 2A)$ for $m_C \gg 2m_A$, and for $\frac{m_C - 2m_A}{m_C} \ll 1$.

Give the total width $\Gamma(C)$ if $m_C > 2m_B$ too. Also express the expected C-lifetime $\langle \tau \rangle$ in terms of $\Gamma(C)$.

(b) Another possible process is the elastic scattering of two particles A and B, schematically

Write down the single diagram that contributes to this process to lowest order. Derive expressions both for the matrix element $_{\text{out}}\langle p_3 p_4 | p_1 p_2\rangle_{\text{in}}$ and for the differential cross section $d\sigma$ ($AB \to AB$).

In the center of mass (CM) frame the process looks like this:

In this frame the differential cross section only depends on the external momenta through $p = |\vec{p}_1|$ and $\theta = \angle(\vec{p}_1, \vec{p}_3)$. Work out $\frac{d\sigma}{d\Omega}(p, \theta)$ in case of equal masses $m_A = m_B \equiv m$. To this purpose, first prove these intermediate steps:

(i) $\int \frac{d^3\bar{p}_3}{E_3} \frac{d^3\bar{p}_4}{E_4} \delta_4(p_3 + p_4 - p_1 - p_2) \overset{\mathrm{CM}}{=} \frac{p}{2\sqrt{p^2+m^2}} \int d\Omega.$

(ii) $[(p_1 \cdot p_2)^2 - m_1^2 m_2^2]^{1/2} \overset{\mathrm{CM}}{=} 2p\sqrt{p^2 + m^2}.$

Calculate the total cross section $\sigma(AB \to AB)$ by integrating over all directions Ω.

Show that in the limit $m_C \to \infty$, while keeping $\lambda \equiv g_{AB}g_B/m_C^2$ constant, $\sigma(AB \to AB)$ is the same as for a model with only A and B particles and interaction

$$\mathcal{L}_{\mathrm{int}} = -\frac{\lambda}{4} A^2 B^2.$$

19. **Lorentz transformation for spinors**

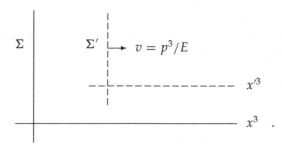

An electron is observed in a frame Σ, where it has velocity v along the 3-axis. Its rest frame is called Σ'. In Σ' the electron's wave function is given by

$$\psi'(x') = \frac{1}{\sqrt{2V'}} \begin{pmatrix} 1 \\ 0 \\ -1 \\ 0 \end{pmatrix} e^{-imt'}. \quad \text{(Weyl representation)}$$

(Due to the volume factor V', ψ' has a volume-independent norm and one can take $V' \to \infty$.)

(a) Verify that this is indeed a positive energy, zero momentum solution of the Dirac equation. What is its spin? Transform the solution to the 'conventional representation' of the lecture notes p. 75.

(b) The wave function ψ in the Σ frame can be determined via a Lorentz transformation. Show that the transformation $K = (K^\mu{}_\nu)$ from coordinates on Σ to coordinates on Σ' [i.e., $(x')^\mu = K^\mu{}_\nu x^\nu$] can be written as

$$K = e^{-\alpha L^{03}} \quad \text{with}$$

$$L^{03} = \begin{pmatrix} 0 & 0 & 0 & 1 \\ 0 & 0 & 0 & 0 \\ 0 & 0 & 0 & 0 \\ 1 & 0 & 0 & 0 \end{pmatrix}, \quad \sinh(\alpha) = \frac{p^3}{m}, \quad \cosh(\alpha) = \frac{E}{m}.$$

(c) Show that the induced transformation of spinors is given by

$$S = \cosh\left(\frac{\alpha}{2}\right)\mathbf{1} + i \sinh\left(\frac{\alpha}{2}\right)\sigma^{03}$$

with $\sigma^{\mu\nu}$ as in the lecture notes (p. 77).

(d) Determine $\psi(x) = S^{-1}\psi'(Kx)$ in the Weyl representation. Verify explicitly in the Σ frame that this is a solution of the Dirac equation with the correct momentum. Finally transform ψ to the conventional representation.

20. **Lorentz algebra vs. su(2)×su(2)**

Using the property that the (Euclidean) Lorentz algebra is isomorphic to su(2)×su(2) one can easily classify all its finite dimensional representations [su(2) is the Lie algebra of SU(2)]. We will analyse this situation in the present exercise.

(a) Show that the matrices $L^{\mu\nu}$ defined by

$$(L^{\mu\nu})^\alpha{}_\beta = g^{\mu\alpha}g^\nu_\beta - g^{\nu\alpha}g^\mu_\beta,$$

generate the Lorentz group (compare this to part (b) of the previous exercise). Furthermore prove that

$$[L^{\mu\nu}, L^{\rho\sigma}] = g^{\nu\rho}L^{\mu\sigma} + g^{\mu\sigma}L^{\nu\rho} - g^{\mu\rho}L^{\nu\sigma} - g^{\nu\sigma}L^{\mu\rho}.$$

(b) Define

$$J_\ell^\pm = \tfrac{1}{2}(\tfrac{1}{2}\varepsilon_{\ell j k}L^{jk} \pm i L_{\ell 0}). \quad (\varepsilon_{123} \equiv +1).$$

Determine all commutators $[J_i^\pm, J_j^\pm]$ and conclude that the Euclidean Lorentz algebra is isomorphic to su(2) × su(2).

(c) It is well known that the set of all finite dimensional representations of su(2) $\cong$ so(3) is given by $\{\rho_l | l = 0, \tfrac{1}{2}, 1, \tfrac{3}{2}, \ldots\}$, where

$$\rho_0(J_i) = 0$$
$$\rho_{\frac{1}{2}}(J_i) = -\tfrac{i}{2}\sigma_i \in su(2)$$
$$\rho_1(J_i) = L_i \in so(3)$$
$$\cdots$$

For each pair (a, b) with $a, b \in \{0, \frac{1}{2}, 1, \frac{3}{2}, \ldots\}$ an irreducible representation of the Euclidean Lorentz algebra can now be defined:

$$\rho_{(a,b)} \equiv \rho_a \otimes \rho_b.$$

In particular $\rho \equiv \rho_{(\frac{1}{2},0)}$ and $\bar{\rho} \equiv \rho_{(0,\frac{1}{2})}$ are defined through

$$\rho(J_i^-)\psi_1 = -\tfrac{i}{2}\sigma_i\psi_1$$
$$\rho(J_i^+)\psi_1 = 0$$
$$\bar{\rho}(J_i^-)\psi_2 = 0$$
$$\bar{\rho}(J_i^+)\psi_2 = -\tfrac{i}{2}\sigma_i\psi_2.$$

Subsequently we can construct the (reducible) representation $\rho \oplus \bar{\rho}$ acting on pairs (ψ_1, ψ_2).

Give the action of $(\rho \oplus \bar{\rho})(J_i^{\pm})$ on (ψ_1, ψ_2).

(d) Now derive the action of $(\rho \oplus \bar{\rho})(L^{\mu\nu})$ on (ψ_1, ψ_2) and note that these objects are precisely the generators $-\tfrac{i}{2}\sigma^{\mu\nu}$ in the Weyl representation (lecture notes p. 77).

21. **γ algebra**

The defining property of the γ matrices $\gamma^1 \cdots \gamma^4$ is

$$\{\gamma^\mu, \gamma^\nu\} = 2g^{\mu\nu}\mathbf{1} \quad \mu, \nu = 0, 1, 2, 3.$$

Furthermore one defines $\gamma^5 \equiv i\gamma^0\gamma^1\gamma^2\gamma^3$.

(a) Show that

$$\{\gamma^\mu, \gamma^5\} = \mathbf{0}, \quad (\gamma^5)^2 = \mathbf{1}.$$

(b) Let $\mathrm{Tr}(\gamma^{\mu_1}\gamma^{\mu_2} \cdots \gamma^{\mu_n})$ denote the trace over n γ-matrices (take $\mu_i \in \{0, 1, 2, 3\}$).

1. Prove that such a trace equals zero for n odd. Also prove that $\mathrm{Tr}\gamma^5 = 0$.

2. Compute

$$\mathrm{Tr}(\gamma^\mu\gamma^\nu)$$
$$\mathrm{Tr}(\gamma^\mu\gamma^\nu\gamma^5)$$
$$\mathrm{Tr}(\gamma^\mu\gamma^\nu\gamma^\rho\gamma^\sigma)$$
$$\mathrm{Tr}(\gamma^\mu\gamma^\nu\gamma^\rho\gamma^\sigma\gamma^5).$$

(c) Prove the following identities:

$$\gamma^\alpha\gamma^\mu\gamma^\nu\gamma^\rho\gamma_\alpha = -2\gamma^\rho\gamma^\nu\gamma^\mu$$
$$\gamma^\mu\gamma^\nu\gamma^\rho + \gamma^\rho\gamma^\nu\gamma^\mu = \tfrac{1}{2}\mathrm{Tr}(\gamma^\alpha\gamma^\mu\gamma^\nu\gamma^\rho)\gamma_\alpha.$$

22. Majorana and Weyl fermions

(a) Given any set of gamma matrices $\{\gamma^\mu\}$ another set is defined by

$$\gamma'^\mu = U^\dagger \gamma^\mu U, \quad U^\dagger U = 1.$$

1. Take $\{\gamma^\mu\}$ to be

$$\gamma^0 = \begin{pmatrix} 1 & 0 \\ 0 & -1 \end{pmatrix}, \quad \gamma^i = \begin{pmatrix} 0 & \sigma_i \\ -\sigma_i & 0 \end{pmatrix}$$

(the 'conventional representation' on pp. 75, 78 of the lecture notes) and choose

$$U = \sigma_1 \otimes \tfrac{1}{2}(\sigma_1 - i\sigma_3) + \sigma_3 \otimes \tfrac{1}{2}(\sigma_1 + i\sigma_3)$$

$$= \frac{1}{2} \begin{pmatrix} \sigma_1 + i\sigma_3 & \sigma_1 - i\sigma_3 \\ \sigma_1 - i\sigma_3 & -(\sigma_1 + i\sigma_3) \end{pmatrix}.$$

Show that U is unitary and that the set $\{\gamma'^\mu\}$ is given by

$$\gamma'^0 = \sigma_3 \otimes \sigma_2, \quad \gamma'^1 = -i \otimes \sigma_3, \quad \gamma'^2 = i\sigma_2 \otimes \sigma_2, \quad \gamma'^3 = -i \otimes \sigma_1.$$

Note that all γ' matrices are purely imaginary.
2. This so-called Majorana representation $\{\gamma'^\mu\}$ allows us to impose the following:

$$\psi^* = \psi \quad \text{(Majorana condition)}$$

[with $\psi = \psi(x)$]. Show that this is consistent with the Dirac equation.
3. Prove that the condition implies $\bar{\psi}\psi = 0$.

Remark: This result is no longer valid for anticommuting $\psi, \bar{\psi}$.
4. How can we interpret Majorana fermions? (charge; antiparticles?)

(b) Another possible condition on fermions is

$$\gamma^5 \psi = \psi. \quad \text{(Weyl condition)}$$

1. Use the Dirac equation to prove that necessarily $m = 0$.
2. Prove that Weyl fermions ω and ψ satisfy $\bar{\omega}\psi = 0$.
3. The helicity operator $\vec{\Sigma} \cdot \hat{k}$ (with $\hat{k} = \vec{k}/|\vec{k}|$) is defined via $\Sigma^i \equiv \tfrac{1}{2}\varepsilon_{ijk}\sigma^{jk}$. Show that in the original γ representation of part (a)(1) this reads

$$\Sigma^i = \begin{pmatrix} \sigma_i & 0 \\ 0 & \sigma_i \end{pmatrix}.$$

Also prove that $(\vec{\Sigma} \cdot \hat{k})^2 = 1$.

4. Show that for $m = 0$ the plane wave solutions of the Dirac equation can be written as

$$k_0 = +E: \quad U_+(\vec{k}) = \begin{pmatrix} \chi \\ (\vec{\sigma} \cdot \hat{k})\chi \end{pmatrix}$$

$$k_0 = -E: \quad U_-(\vec{k}) = \begin{pmatrix} -(\vec{\sigma} \cdot \hat{k})\chi \\ \chi \end{pmatrix}.$$

Determine the action on $U_\pm$ of the chirality operator γ^5 and the helicity operator $\vec{\Sigma} \cdot \hat{k}$. In particular show that their action is the same, up to signs.

5. Conclude that a massless spinor satisfying the Weyl condition can either describe right-handed particles or left-handed antiparticles. Here right- (left) handed means having positive (negative) helicity.

Remark: From (b)(3) it is clear that helicity equals the (anti-) particle's spin component in its direction of movement. Therefore the helicity operator commutes with the Dirac equation for any mass m, this equation being Lorentz (hence rotation) covariant. Chirality, however, is only a good quantum number in the massless case.

(c) Is it possible to realise the Majorana and Weyl conditions simultaneously?

23. **Dirac equation**

We start from the Dirac action

$$S_{\text{Dirac}} = \int d^4x\, \bar{\psi}(x)(i\gamma^\mu \partial_\mu - m)\psi(x).$$

(a) In the lecture notes (p. 14) the energy-momentum tensor $T^{\mu\nu}$ has been constructed for the (scalar) bosonic case. As this construction uses general coordinate invariance, its generalisation to fermions is complicated (the formulation of spinors in general relativity is involved). For this we refer to Section 10 of 'The Spacetime Approach to Quantum Field Theory,' by B.S. DeWitt in *Relativity, Groups and Topology II*, ed. B.S. DeWitt and R. Stora (North-Holland, Amsterdam, 1984). The generalisation of eq. (3.20) yields an energy-momentum tensor that is no longer symmetric:

$$T^{\mu\nu} = \bar{\psi}i\gamma^\mu\partial^\nu\psi - g^{\mu\nu}\bar{\psi}(i\gamma^\alpha\partial_\alpha - m)\psi$$
$$+ \frac{1}{4}\partial_\lambda\left(\bar{\psi}(\gamma^\lambda\sigma^{\mu\nu} - \gamma^\mu\sigma^{\lambda\nu} + \gamma^\nu\sigma^{\mu\lambda})\psi\right).$$

Use the equations of motion to show that $\partial_\mu T^{\mu\nu} = 0$ and that the energy-momentum tensor is equivalent to the following symmetric result:

$$T^{\mu\nu} = \frac{1}{4}\left(\bar\psi i\gamma^\mu\partial^\nu\psi + \bar\psi i\gamma^\nu\partial^\mu\psi - (\partial^\nu\bar\psi)i\gamma^\mu\psi - (\partial^\mu\bar\psi)i\gamma^\nu\psi\right).$$

It is possible, however, to use only translational invariance (and of course, as always for Noether currents, the equations of motion). This derivation is very close in spirit to the discussion on pg. 14 and to eq. (3.20). Show that translational invariance can be formulated in the following way:

$$\mathcal{L}[\psi_\Lambda, \bar\psi_\Lambda, \partial_\mu\psi_\Lambda, \partial_\mu\bar\psi_\Lambda](x - \Lambda) = \mathcal{L}[\psi, \bar\psi, \partial_\mu\psi, \partial_\mu\bar\psi](x),$$

$$(23.10)$$

where $\psi_\Lambda(x) \equiv \psi(x + \Lambda)$ (Λ independent of x), etc.

Expand the left-hand side of eq. (23.10) to first order in Λ. Now use the equations of motion, and eq. (23.10), to prove that $\partial_\mu T^{\mu\nu} = 0$ for

$$T^{\mu\nu} = \bar\psi i\gamma^\mu\partial^\nu\psi - g^{\mu\nu}\bar\psi(i\gamma^\alpha\partial_\alpha - m)\psi.$$

Also verify *explicitly* from the Dirac equation that $\partial_\mu T^{\mu\nu} = 0$ is satisfied.

Show that all three definitions of the energy-momentum tensor give the same results for $H = \int d_3\vec{x}\, T_{00}$ and $P_i = \int d_3\vec{x}\, T_{0i}$, that these quantities are conserved and that for a plane wave solution of the Dirac equation with momentum $\vec{k}$ (see Chapter 13) they coincide, as it should be, with the energy and momentum of that solution, i.e., $H = k_0(\vec{k})$ and $P_i = k_i$.

(b) Now add interactions and (external) sources:

$$\mathcal{L} = \mathcal{L}_{\text{Dirac}} - V_{\text{int}} + \mathcal{L}_{\text{source}}, \quad V_{\text{int}} = \tfrac{1}{4}g(\bar\psi\psi)^2,$$
$$\mathcal{L}_{\text{source}} = -(\bar J\psi + J\bar\psi).$$

What are the corresponding Euler–Lagrange equations for ψ and $\bar\psi$? Solve these equations in a perturbative way, like in the scalar case (Problem 7; pp. 19–22 of the lecture notes). In particular give the Feynman rules for the equivalent diagrammatic expansion. For the lowest order result use the following notation:

$$\psi = \underset{\bar J}{\underrightarrow{\quad\quad}}\!\!\!\times \quad ,$$

$$\bar\psi = \underset{J}{\underrightarrow{\quad\quad}}\!\!\!\times \quad .$$

24. **Canonical formalism for spinors**
    (a) 1. On p. 84 of the lecture notes creation and annihilation operators for the Dirac field are introduced:

$$\psi(x) = \int \frac{d^3\vec{k}}{(2\pi)^{3/2}} \frac{1}{\sqrt{2k_0}} \sum_{a=1}^{2} \left(b_a(\vec{k})u^{(a)}(\vec{k})e^{-ikx} + d_a^\dagger(\vec{k})v^{(a)}(\vec{k})e^{ikx} \right),$$

with $k_0 = +\sqrt{\vec{k}^2 + m^2}$. Give the corresponding expression for $\bar{\psi}$.

2. Postulate anticommutation relations as on p. 86:

$$\{b_a(\vec{k}), b_b^\dagger(\vec{k}')\} = \{d_a(\vec{k}), d_b^\dagger(\vec{k}')\} = \delta_{ab}\delta^3(\vec{k} - \vec{k}');$$

the remaining anticommutators are zero. Show that

$$\{\psi_\alpha(x), \psi_\beta(y)\} = \{\bar{\psi}_\alpha(x), \bar{\psi}_\beta(y)\} = 0 \quad \text{and}$$
$$\{\psi_\alpha(x), \bar{\psi}_\beta(y)\} = (i\gamma^\mu \partial_\mu + m)_{\alpha\beta}\Delta(x - y), \quad \text{with}$$
$$\Delta(x - y) = \int \frac{d^3\vec{k}}{(2\pi)^3} \frac{1}{2k_0} \left(e^{-ik(x-y)} - e^{ik(x-y)} \right)$$

(and $\partial_\mu = \partial/\partial x^\mu$). Compare with Problem 6 and conclude that causality is respected.

3. Alternatively, postulate commutation relations (substitute $\{\, , \,\} \rightarrow [\, , \,]$ in above anticommutation relations). Show that

$$[\psi_\alpha(x), \psi_\beta(y)] = [\bar{\psi}_\alpha(x), \bar{\psi}_\beta(y)] = 0 \quad \text{and}$$
$$[\psi_\alpha(x), \bar{\psi}_\beta(y)] = (i\gamma^\mu \partial_\mu + m)_{\alpha\beta}\tilde{\Delta}(x - y), \quad \text{with}$$
$$\tilde{\Delta}(x - y) = \int \frac{d^3\vec{k}}{(2\pi)^3} \frac{1}{2k_0} \left(e^{-ik(x-y)} + e^{ik(x-y)} \right).$$

Conclude that causality is violated.

Remark: This result can be generalised to a theorem stating that any local quantum field theory that respects causality admits only fermions with half integer spin and bosons with integer spin.

(b) Add a source term $\bar{J}\psi + \bar{\psi}J$ to the Dirac Hamiltonian density $\bar{\psi}(-i\gamma^i\partial_i + m)\psi$, $J_\alpha(x)$ and $\bar{J}_\alpha(x)$ being anticommuting external fields. Expand $\langle 0| \exp(-iHt)|0\rangle$ up to second order in the sources. To this purpose use the Hamiltonian perturbation formalism (lecture notes pp. 25–28); use the properties of u, v (p. 84) and the gamma matrices to simplify the spinor structure. Your final result should be

$$\langle 0|e^{-iHt}|0\rangle e^{iE_0 t} = 1 - i \int d^4x d^4y \bar{J}(x) G_F(x-y) J(y) + \cdots \quad \text{with}$$

$$G_F(x-y) = \int \frac{d^4k}{(2\pi)^4} \frac{k_\mu \gamma^\mu + m}{k^2 - m^2 + i\varepsilon} e^{-ik(x-y)}.$$

Note that G_F is precisely the classical fermion propagator of the previous exercise.

25. **Anticommuting variables**

In this exercise Greek letters denote anticommuting variables, ordinary letters commuting ones.

(a) Compute the following integrals:

$$\int d\theta\, e^{\theta a}, \quad \int d\theta\, \frac{1}{1-a\theta}, \quad \int d\theta\, \ln(1+\theta).$$

(b) Given the following linear relation between two sets of n-independent anticommuting variables,

$$\eta_i = \sum_{j=1}^n B_{ij}\theta_j,$$

show that (for invertible B)

$$d\eta_1 d\eta_2 \cdots d\eta_n = \frac{1}{\det B} d\theta_1 d\theta_2 \cdots d\theta_n.$$

Compare this to the case of commuting variables.

Hint: Consider the most general function of n anticommuting variables, which is a polynomial of degree n. Analyse its behaviour under integrations and linear transformations on the variables.

(c) Prove that, for independent $\eta_i, \bar{\eta}_j$,

$$\int d\eta_1 d\bar{\eta}_1 \cdots d\eta_n d\bar{\eta}_n e^{\bar{\eta}_i A_{ij} \eta_j} = \det A.$$

Use this result to prove the following result, which holds for any antisymmetric matrix A:

$$\int d\theta_1 \cdots d\theta_n e^{\frac{1}{2}\theta_i A_{ij}\theta_j} = \pm\sqrt{\det A}.$$

Hint: Substitute $\eta_i = \theta_i + i\bar\theta_i$, $\bar\eta_i = \theta_i - i\bar\theta_i$.

(d) Given a smooth function f satisfying $\lim_{y\to\infty} f(y) = 0$, prove that

$$\int dx_1 dx_2 d\theta d\bar\theta\, f(x_1^2 + x_2^2 + \bar\theta\theta) = -\pi f(0).$$

26. **One-loop Feynman diagrams**

Consider a model consisting of fermions ψ and real scalar particles φ with interaction

$$V_{\text{int}} = g\bar\psi\varphi\psi.$$

Determine the reduced matrix elements corresponding to the following diagrams (do <u>not</u> work out the analytical expressions):

1. φ self-energy

2. ψ self-energy

3. vertex correction

27. **Compton scattering for pions**

At not-too-high energies the pion–photon interaction is well approximated by scalar QED:

$$\mathcal{L} = -\tfrac{1}{4}F_{\mu\nu}F^{\mu\nu} + (\partial_\mu - ie A_\mu)\varphi^*(\partial^\mu + ie A^\mu)\varphi - m^2\varphi^*\varphi.$$

The pion (π^-) is described by the complex scalar field φ, the photon (γ) by the vector field A_μ ($F_{\mu\nu} = \partial_\mu A_\nu - \partial_\nu A_\mu$).

(a) Show that the three-point vertex is given by

$$= e(p_\mu + q_\mu),$$

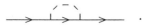

and give the other Feynman rules in the Lorentz gauge.

(b) Which Feynman diagrams contribute, to order e^2, to $\pi^-\gamma \to \pi^-\gamma$ elastic scattering?

(c) The initial and final photon states are plane waves:

$$\frac{\varepsilon_\mu^{\text{in}}(\vec k)e^{-ik^i x_i}}{\sqrt{(2\pi)^3 2k_0}} \quad \text{resp.} \quad \frac{\varepsilon_\mu^{\text{out}}(\vec k')e^{-ik'^i x_i}}{\sqrt{(2\pi)^3 2k_0'}}.$$

Express the reduced matrix element for the scattering process, to order e^2, in terms of the polarisation vectors $\varepsilon^{\text{in,out}}$ and the external momenta. Use the following notation:

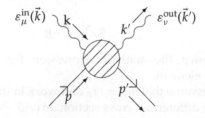

(d) Using the result of part (c) prove (to order e^2) that the S-matrix vanishes whenever the initial or final photon is longitudinal (i.e., $\varepsilon_\mu(\vec{k}) \sim k_\mu$). Explain which property of the model is responsible for this.

(e) Give all Feynman diagrams that are needed for an order e^6 calculation of the *cross section*. Which of them are *UV* divergent, i.e., which give rise to expressions that diverge due to integrations over large momenta?

28. **Elementary fermionic processes**
   Let us reconsider the situation in Problem 18, with the bosonic fields A and B replaced by fermionic fields ψ_A and ψ_B. The Lagrangian now is

   $$\mathcal{L} = \tfrac{1}{2}\partial_\mu C \partial^\mu C - \tfrac{1}{2}m_C^2 C^2 + \bar{\psi}_A(i\,\partial\!\!\!/ - m_A)\psi_A + \bar{\psi}_B(i\,\partial\!\!\!/ - m_B)\psi_B$$
   $$-g_A\bar{\psi}_A C\psi_A - g_B\bar{\psi}_B C\psi_B.$$

   (a) For $m_C > 2m_A$, C can decay into A and anti-A according to the diagram

   Determine like in Problem 18(a) the S-matrix element and the decay width Γ, which now are functions of the fermion spins. Perform a summation over all possible spins to obtain the expected lifetime of C.

   <u>Hint</u>: Some properties of the u and v spinors (lecture notes p. 84) are very useful here.

(b) Scattering of A and B is described by

Write down the analytic expression for the corresponding S-matrix element.

Now assume that $m_A = m_B$ and work in the CM frame. Determine the differential cross section $d\sigma(AB \to AB)$. *Average* over the incoming spins and *sum* over the outgoing spins. For which experimental situation is this justified?

Work out your result as a function of the CM variables $|\vec{p}|$ and θ. As a check it is given that $d\sigma/d\Omega$ is spherically symmetric for $m_C = 2m_A$.

29. e^-e^+ **collisions in QED**

The QED Lagrangian with two flavours, electrons and muons, reads

$$\mathcal{L} = -\tfrac{1}{4}F_{\mu\nu}F^{\mu\nu} - \tfrac{1}{2}\alpha_{\text{gauge}}(\partial_\mu A^\mu)^2 + \sum_{f=e,\mu} \bar{\psi}_f(i\gamma^\mu D_\mu - m_f)\psi_f,$$

where $D_\mu = \partial_\mu - ieA_\mu$ (lecture notes p. 113).

(a) $e^-e^+ \to e^-e^+$ (Bhabha scattering)

Which two diagrams contribute to lowest order?

In the lecture notes the Møller $(e^-e^- \to e^-e^-)$ differential cross section is calculated. Perform an analogous calculation to obtain (in the CM frame)

$$\frac{d\sigma}{d\Omega}(e^-e^+ \to e^-e^+)$$

$$= \frac{\alpha^2}{16E^2} \left\{ \frac{(2E^2 - m^2)^2}{(E^2 - m^2)^2 \sin^4(\theta/2)} + \frac{-8E^4 + m^4}{E^2(E^2 - m^2)\sin^2(\theta/2)} \right.$$

$$+ \frac{12E^4 + m^4}{E^4} - \frac{4(2E^2 - m^2)(E^2 - m^2)\sin^2(\theta/2)}{E^4}$$

$$\left. + \frac{4(E^2 - m^2)^2 \sin^4(\theta/2)}{E^4} \right\},$$

where $\alpha = e^2/4\pi$, $m = m_e$; E and θ are the CM variables for the energy of the incoming electron resp. the angle between the in- and outgoing electrons.

Note that unlike in the Møller case there is no divergence at $\theta = \pi$. What is the reason for this?

(b) $e^-e^+ \to \mu^-\mu^+$

How many diagrams contribute to lowest order?

Show that in the CM frame, and in the limit $m_e/E, m_\mu/E \to 0$,

$$\frac{d\sigma}{d\Omega}(e^-e^+ \to \mu^-\mu^+) = \frac{\alpha^2}{16E^2}(1 + \cos^2\theta).$$

Calculate the total cross section. Use dimensional analysis to express your result in units $\hbar, c \neq 1$.

30. **Weak interaction in the standard model**

The Lagrangian given below describes a simplified version of the standard model. This simplification, which only contains fermions ψ and massive vector bosons W_μ, captures the mechanism through which the standard model gives rise to an effective four-fermion interaction.

$$\mathcal{L}(W_\mu, \psi) = -\tfrac{1}{4}F_{\mu\nu}F^{\mu\nu} + \tfrac{1}{2}M^2 W_\mu W^\mu + \bar{\psi}(i\gamma^\mu\partial_\mu - m)\psi + gW_\mu(\bar{\psi}\gamma^\mu\psi)$$

with $F_{\mu\nu} = \partial_\mu W_\nu - \partial_\nu W_\mu$.

(a) Give the Feynman rules.

(b) We restrict ourselves to tree diagrams which satisfy two conditions: (1) all external lines are fermionic; (2) $p^2 \ll M^2$ for all momenta p_μ. Show that such diagrams can effectively be described by

$$\mathcal{L}_{\text{effective}}(\psi) = \bar{\psi}(i\gamma^\mu\partial_\mu - m)\psi - \frac{\lambda}{(2!)^2}(\bar{\psi}\gamma^\mu\psi)(\bar{\psi}\gamma_\mu\psi)$$

and express the parameter λ in terms of g and M.

31. **Gauge fields**

In this exercise we use the following gauge-field conventions (compare this to the lecture notes):

$$A_\mu = q A_\mu^a T^a, \quad F_{\mu\nu} = q F_{\mu\nu}^a T^a, \quad [T^a, T^b] = f_{abc} T^c,$$
$$\text{Tr}(T^a T^b) = -\tfrac{1}{2}\delta_{ab}, \quad D_\mu = \partial_\mu + A_\mu,$$
$$F_{\mu\nu} = [D_\mu, D_\nu]$$
$$= \partial_\mu A_\nu - \partial_\nu A_\mu + [A_\mu, A_\nu],$$

where $\{T^a\}$ spans a matrix representation $\rho(L_G)$ of the Lie algebra. Note that we absorb the coupling q in A_μ.

(a) <u>Notation</u>: X, Y, Z stand for arbitrary elements of the Lie algebra $\rho(L_G)$.

In the lecture notes the generators of the adjoint representation are defined as

$$(\text{ad}T^a)_{bc} = -f_{abc}.$$

Show that this representation can be thought of as acting on the Lie algebra itself, in the following way:

$$(\mathrm{ad}T^a)Y = [T^a, Y].$$

Also prove from this formula that $X \rightarrow \mathrm{ad}X$ indeed is a representation of the Lie algebra, i.e., prove that it is a linear map, satisfying

$$(\mathrm{ad}[X, Y]) = [(\mathrm{ad}X), (\mathrm{ad}Y)].$$

Hint: Work out $(\mathrm{ad}[X, Y])Z$, using the Jacobi identity $[X, [Y, Z]]$ + cyclic $= 0$ (which can be seen to hold trivially by expanding the commutators).
Finally prove that

$$e^{\mathrm{ad}X}Y = e^X Y e^{-X}.$$

This means that the adjoint representation of the *group* G is a conjugation. Therefore gauge transformations act on the field strength through the adjoint group representation (as $F_{\mu\nu} \rightarrow g F_{\mu\nu} g^{-1}$, lecture notes p. 131).

(b) Define $D_\mu^{(\mathrm{ad})} X \equiv (\mathrm{ad}D_\mu)X = [\partial_\mu + A_\mu, X] = (\partial_\mu X) + [A_\mu, X]$. Prove that

$$D_\mu^{(\mathrm{ad})} D_\nu^{(\mathrm{ad})} F^{\mu\nu} = 0.$$

Hint: What is $[D_\mu^{(\mathrm{ad})}, D_\nu^{(\mathrm{ad})}]$?

(c) The gauge-invariant Lagrangian for a fermion field coupled to a dynamical SU(N) gauge field is (lecture notes pp. 130, 131)

$$\mathcal{L} = \bar{\psi}(i\gamma^\mu D_\mu - m)\psi + \frac{1}{2q^2}\mathrm{Tr}(F_{\mu\nu}F^{\mu\nu}) =$$

$$= \bar{\psi}(i\gamma^\mu \partial_\mu - m)\psi + \frac{1}{2q^2}\mathrm{Tr}(F_{\mu\nu}F^{\mu\nu}) + \frac{2}{q^2}\mathrm{Tr}(J^\mu A_\mu) \quad \text{with}$$

$$J^\mu = qJ^{\mu a}T^a, \quad J^{\mu a} = -iq\bar{\psi}\gamma^\mu T^a \psi.$$

Derive the Euler–Lagrange equations:

$$D_\mu^{(\mathrm{ad})} F^{\mu\nu} = J^\nu \tag{23.11}$$

$$(i\gamma^\mu D_\mu - m)\psi = 0 \tag{23.12}$$

Show from eq. (23.11) that

$$D_\mu^{(\mathrm{ad})} J^\mu = 0.$$

Show that this equation follows from eq. (23.12) as well.

(d) Use the Jacobi identity to prove the Bianchi identity:

$$D_\mu^{(ad)} F_{\nu\rho} + \text{cyclic} = 0.$$

Show that for electromagnetism [$G =$ U(1)] this gives the homogeneous Maxwell equations.

32. **Dirac equation with gauge fields**

(a) By construction the Klein–Gordon equation is obtained from the free Dirac equation in the following way:

$$0 = -(i\gamma^\mu \partial_\mu + m)(i\gamma^\nu \partial_\nu - m)\psi = (\partial^2 + m^2)\psi.$$

Analogously prove from the Yang–Mills Dirac equation [Problem 31(c) eq. (23.12)] that

$$(D^2 + m^2 - \tfrac{i}{2}\sigma^{\mu\nu} F_{\mu\nu})\psi = 0. \tag{23.13}$$

(b) Now specify to electromagnetism,

$$T = i,\ q = -e,\ F_{\mu\nu} = \partial_\mu A_\nu - \partial_\nu A_\mu, \quad \text{etc.}$$

Choose the gauge $A_0 = 0$ and turn off the electric field by assuming that $\partial_t \vec{A} = \vec{0}$. Show that eq. (23.13) reduces to

$$(D^2 + m^2)\psi + e \begin{pmatrix} \vec{\sigma} \cdot \vec{B} & 0 \\ 0 & \vec{\sigma} \cdot \vec{B} \end{pmatrix} \psi = 0.$$

Write $\psi = \begin{pmatrix} \psi_G \\ \psi_S \end{pmatrix}$ and define the two-spinor ψ_{sch} to be

$$\psi_{\text{sch}} = e^{imt}\psi_G$$

(subtraction of the rest energy from the Hamiltonian). Show that[1]

$$-\frac{1}{2m}\frac{\partial^2}{\partial t^2}\psi_{\text{sch}} + i\frac{\partial}{\partial t}\psi_{\text{sch}} = \left[\frac{(\vec{p}+e\vec{A})^2}{2m} + \frac{e}{2m}(\vec{\sigma}\cdot\vec{B})\right]\psi_{\text{sch}}.$$

Show that in the nonrelativistic limit this equation simplifies to the well-known Schrödinger equation for an electron in a magnetic field, i.e.,

$$\begin{cases} i\dfrac{\partial}{\partial t}\psi_{\text{sch}} = H\psi_{\text{sch}} \\ H = \dfrac{1}{2m}(\vec{p}+e\vec{A})^2 + \dfrac{e}{2m}(\vec{\sigma}\cdot\vec{B}). \end{cases}$$

---

[1] Please note: To respect covariance $p_k = -i\partial/\partial x^k$, whereas $p^k = -i\partial/\partial x^k$ or $\vec{p} = (p^1, p^2, p^3) = -i\vec{\nabla}$ as used in nonrelativistic quantum mechanics.

33. **Linear sigma model**

The Lagrangian for the linear sigma model reads

$$\mathcal{L} = \tfrac{1}{2}[\partial_\mu \vec{\varphi} \cdot \partial^\mu \vec{\varphi} + \partial_\mu \sigma \partial^\mu \sigma] + \tfrac{1}{4}\mu^2[|\vec{\varphi}|^2 + (\sigma + v)^2]$$
$$- g[|\vec{\varphi}|^2 + (\sigma + v)^2]^2,$$

where $\vec{\varphi}$ is the pion field (three real components), σ is the sigma field (one real component), v a constant and μ, g are real positive parameters.

(a) Show that this Lagrangian contains a linear term $\alpha\sigma$ and express α in terms of μ, g and v. What is the Feynman rule for such a term?

For $\alpha \neq 0$, this Feynman rule makes the perturbative approach unnecessarily complicated. Argue that this complication is avoided when v is such that $\vec{\varphi} = \vec{0}$ and $\sigma = 0$ corresponds to the minumum of the potential associated to the Lagrangian.

Determine v, and show that the $\vec{\varphi}$ and σ masses are 0 resp. μ.

(b) Show that the Lagrangian is invariant under the global infinitesimal (isospin) transformations

$$\delta_\Lambda \sigma(x) = 0, \quad \delta_\Lambda \varphi_i(x) = -\varepsilon_{ijk}\Lambda_j \varphi_k(x), \quad (23.14)$$

and also under the global transformations

$$\delta_\xi \sigma(x) = -\vec{\varphi}(x) \cdot \vec{\xi}, \quad \delta_\xi \vec{\varphi}(x) = (\sigma(x) + v)\vec{\xi}. \quad (23.15)$$

Now write eqs. (23.14) and (23.15) in matrix notation w.r.t. the four-component vector $\varphi_\mu(x)$ by defining $\varphi_4(x) \equiv \sigma(x) + v$ (i.e., write $\delta_\Lambda \varphi_\mu(x) = \Lambda_i L^i_{\mu\nu}\varphi_\nu(x)$, w.r.t. $\delta_\xi \varphi_\mu(x) = \xi_i K^i_{\mu\nu}\varphi_\nu(x)$ for suitably defined 4×4 matrices L^i and K^i). Prove that L^i and K^i span the space of real antisymmetric 4×4 matrices.

Conclude that eqs. (23.14) and (23.15) are in fact infinitesimal SO(4) transformations. Verify this by showing that the Lagrangian, written in terms of the four-component vector φ_μ, is manifestly SO(4) invariant.

(c) Give the Noether currents associated to eqs. (23.14) and (23.15)

34. **Higgs mechanism**

We consider a model with real scalar fields φ^i and vector fields A^i_μ, $i = 1, 2, 3$. These fields transform in the fundamental representation of an internal group SO(3), with generators

$$(T^i)_{jk} = -\epsilon_{ijk} \quad i, j, k = 1, 2, 3.$$

In particular the covariant derivative and field tensor read

$$(D_\mu \varphi)^i = \partial_\mu \varphi^i - g\epsilon_{kij} A^k_\mu \varphi^j,$$
$$F^i_{\mu\nu} = \partial_\mu A^i_\nu - \partial_\nu A^i_\mu + g\epsilon_{ijk} A^j_\mu A^k_\nu.$$

The Lagrangian is taken to be

$$\mathcal{L} = -\tfrac{1}{4} F^i_{\mu\nu} F^{\mu\nu i} + \tfrac{1}{2} (D_\mu \varphi)^i (D^\mu \varphi)^i - V(|\vec{\varphi}|^2) \text{ with potential}$$
$$V(|\vec{\varphi}|^2) = \tfrac{\lambda}{4} [|\vec{\varphi}|^2 - F^2]^2, \qquad\qquad \lambda, F > 0.$$

The (0-loop) vacuum expectation value (vev) of the scalar fields is chosen to be

$$\langle \vec{\varphi}(x) \rangle = \vec{F} \equiv \begin{pmatrix} 0 \\ 0 \\ F \end{pmatrix}, (F \text{ constant}).$$

(a) Explain why this is a valid choice for the vev. Show that this vev is invariant under a one dimensional subgroup of SO(3).

(b) Define $\tilde{\varphi}^i = \varphi^i - F^i$, and expand the Lagrangian in terms of $\tilde{\varphi}^i$ and A^i_μ. Note that the quadratic part of the Lagrangian contains off-diagonal elements (mixing A and $\tilde{\varphi}$). In general such terms can be handled by diagonalisation (redefining A and $\tilde{\varphi}$ in terms of each other in an appropriate way), but anticipating the gauge choice in part (c) you may neglect them here.

   Interpret the various terms (mass terms, kinetic terms, interaction terms). Give the masses for the fields A^i_μ and $\tilde{\varphi}^i$ ($i = 1,2,3$), and also the couplings for the following three-vertices:

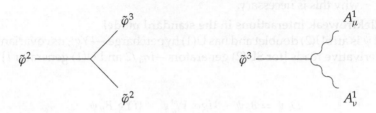

(c) Show that we can choose the gauge such that

$$\tilde{\varphi}^1 = \tilde{\varphi}^2 = 0.$$

Does this completely fix the gauge?

    This model contains nine physical degrees of freedom (dof). Read off from the quadratic terms how in the above gauge these physical dof are distributed over the fields.

(d) Reconsider the situation for a different potential, $V(|\vec{\varphi}|^2) = \frac{1}{2}m^2|\vec{\varphi}|^2$. What is the $\vec{\varphi}$ vev in this case? Read off the number of physical dof again ($\varphi^1 = \varphi^2 = 0$ is <u>not</u> a convenient gauge choice now. Why?).

35. **Higgs effect and ghosts**
Take the model in the previous exercise and add the following gauge-fixing term to the Lagrangian:

$$\mathcal{L}^{\text{gauge}} = -\frac{\alpha}{2}\mathcal{F}_a^2 \,, \quad \mathcal{F}_a = \partial_\mu A^{\mu a} - \frac{g}{\alpha}\epsilon_{abc}F^b\varphi^c .$$

(a) Expand $\mathcal{L} + \mathcal{L}^{\text{gauge}}$ up to quadratic terms in A_μ^a and $\tilde{\varphi}^a \equiv \varphi^a - F^a$. For convenience choose $F^a = F\delta_{a3}$, as in the previous exercise. Show that due to the special gauge choice, the quadratic terms mixing A and $\tilde{\varphi}$ cancel among $\mathcal{L}$ and $\mathcal{L}^{\text{gauge}}$.

(b) Determine how $\mathcal{F}_a$ transforms under infinitesimal local gauge transformations $\varphi \to \Omega\varphi$; $A_\mu \equiv A_\mu^a T^a \to \Omega A_\mu \Omega^{-1} + g^{-1}\Omega\partial_\mu\Omega^{-1}$ with $\Omega_{ij} = (\exp(\Lambda^a T^a))_{ij} = \delta_{ij} - \Lambda^a\epsilon_{aij}$, and write the result as $\delta\mathcal{F}_a = M_{ab}\Lambda^b$ (compare this to lecture notes pp. 137, 138). Read off the ghost masses.

(c) Determine the vector, scalar and ghost propagators as a function of α. Which limits correspond to the transversal gauge ($\partial_\mu A^{\mu a} = 0$) and the unitary gauge ($\tilde{\varphi}^1 = \tilde{\varphi}^2 = 0$, as in part (c) of the previous exercise)?

(d) Which poles in the propagators correspond to *physical* masses? Check that unphysical poles always coincide mutually, and argue why this is necessary.

36. **Elektroweak interactions in the standard model**
If ψ is an SU(2) doublet and has U(1) hypercharge $-\frac{1}{2}Yg'$, its covariant derivative reads [for SU(2) generators $-i\sigma_a/2$ and U(1) generator i]

$$D_\mu\psi = \partial_\mu\psi - \frac{1}{2}ig\sigma_a W_\mu^a\psi - \frac{1}{2}iYg'B_\mu\psi.$$

It is given that the fermion fields e^L, e^R (electron), v (neutrino), u^L, u^R, d^L, d^R (up and down quarks) have the following SU(2)×U(1) properties:

$$\psi_e^L = \begin{pmatrix} \nu \\ e^L \end{pmatrix} \qquad \text{doublet} \quad Y = -1$$

$$\psi_e^R = e^R \qquad \text{singlet} \quad Y = -2$$

$$\psi_q^L = \begin{pmatrix} u^L \\ d^L \end{pmatrix} \qquad \text{doublet} \quad Y = +\tfrac{1}{3}$$

$$\psi_u^R = u^R \qquad \text{singlet} \quad Y = +\tfrac{4}{3}$$

$$\psi_d^R = d^R \qquad \text{singlet} \quad Y = -\tfrac{2}{3}$$

Furthermore, we reformulate the gauge fields:

$$W_\mu^\pm = \frac{1}{\sqrt{2}}(W_\mu^1 \mp i W_\mu^2),$$

$$Z_\mu = -W_\mu^3 \cos\theta_W + B_\mu \sin\theta_W, \qquad A_\mu^{\text{em}} = W_\mu^3 \sin\theta_W + B_\mu \cos\theta_W,$$

and $\sigma_\pm = \sigma_1 \pm i\sigma_2$.

(a) Write the covariant derivative of ψ_e^L and ψ_e^R in terms of $W_\mu^\pm$, Z_μ, A_μ^{em} and $\sigma_+, \sigma_-, \sigma_3$.

(b) We require the interaction between the electron fields e^L, e^R and the photon field A_μ^{em} to be the same as in quantum electrodynamics (QED). Derive from this requirement two relations between g, g', θ_W and the electron charge $-e$.

(c) Work out the relevant covariant derivatives to determine the electromagnetic charge of the neutrino and the up and down quarks. Also analyse the electromagnetic properties of the fields $W_\mu^\pm$ and Z_μ. Discuss the particle interpretation of the complex fields $W_\mu^\pm$.

(d) The Lagrangian of the standard model contains (a.o.) the following terms:

$$i\bar{\psi}_e^L \gamma^\mu D_\mu \psi_e^L + i\bar{\psi}_e^R \gamma^\mu D_\mu \psi_e^R.$$

Determine from this all possible three-vertices of type ⟶⟨ ,

where ∿∿ stands for W_μ^+, W_μ^- or Z_μ, and ⟶ for e^L, e^R or ν.

(e) The standard model allows for the process $W^- \to e^L + \bar{\nu}$. It is given that the decay rate Γ_0 equals

$$\Gamma_0 \equiv \Gamma(W^- \to e^L \bar{\nu}) = \frac{1}{48\pi} g^2 M_W$$

(M_W is the mass of W^-, the masses of e^L, e^R and ν are neglected). Use your results from part (d) to express the decay rates $\Gamma_1 \equiv \Gamma(Z \to e^R \bar{e}^R)$ and $\Gamma_2 \equiv \Gamma(Z \to \nu \bar{\nu})$ in terms of Γ_0 and θ_W.

(f) Show that

$$\Gamma(Z \to e^L \bar{e}^L) \neq \Gamma(Z \to e^R \bar{e}^R)$$

and interpret this result.

37. **LEP experiment**

Since 1989 CERN has been operating the Large Electron–Positron (LEP) collider, a ring with a circumference of 27 km. Electrons (e^-) and positrons (e^+) are accelerated in opposite directions, each reaching an energy (E) of about 45 GeV. The CM energy ($2E$) in a collision is comparable to the mass of the neutral vector boson Z that was encountered in the previous exercise. From Heisenberg's uncertainty relation it is then clear that a Z boson created in the collision can exist for a relatively long time. This gives rise to a so-called resonance in the electron–positron cross section. In the present exercise we will analyse this phenomenon for the process $e^+ e^- \to \mu^+ \mu^-$.

The following part of the standard model Lagrangian (in the unitary gauge) suffices for a leading order calculation:

$$\mathcal{L} = -\tfrac{1}{4}(\partial_\mu Z_\nu - \partial_\nu Z_\mu)(\partial^\mu Z^\nu - \partial^\nu Z^\mu) + \tfrac{1}{2}M_Z^2 Z_\mu Z^\mu +$$
$$+ \sum_{f=e,\mu} \left\{ \bar{\psi}_f (i\gamma^\mu \partial_\mu - m_f)\psi_f - \frac{e}{\sqrt{3}} \bar{\psi}_f Z_\mu \gamma^\mu \gamma_5 \psi_f \right\}.$$

In this Lagrangian the spinor field $\psi_{e(\mu)}$ describes the electron (muon) with mass $m_{e(\mu)} = 0.511$ MeV (105.7 MeV), while the Z particle (with mass $M_Z = 91.2$ GeV) is described by the vector field Z_μ. For convenience the Weinberg angle θ_W has been approximated ($\sin^2 \theta_W = 0.25$ instead of $\sin^2 \theta_W = 0.23$). As can be seen from part (d) of the previous exercise, this considerably simplifies the interaction between the Z particle and the fermions. The Feynman rules now are

$$\mu \overset{k}{\sim\!\sim\!\sim} \nu \quad = \quad \frac{-(g_{\mu\nu} - \frac{k_\mu k_\nu}{M_Z^2})}{k^2 - M_Z^2 + i\epsilon},\qquad b \overset{k}{\longrightarrow} a \quad = \quad \frac{(m_f + \gamma^\mu k_\mu)_{ab}}{k^2 - m_f^2 + i\epsilon},$$

$$b \overset{\mu}{\longrightarrow}\!\!\!<_a \quad = \quad \frac{e}{\sqrt{3}}(\gamma^\mu \gamma_5)_{ab}.$$

(a) Give all Feynman diagrams and the **S** matrix for $e^+e^- \rightarrow \mu^+\mu^-$ via a Z particle. Also give the Feynman diagrams for $e^+e^- \rightarrow e^+e^-$ via a Z particle.

(b) Prove from the Dirac equation that $\bar{u}^{s_1}(p)\gamma^\mu v^{s_2}(q)(p_\mu + q_\mu) = 0$ and the same for $u \leftrightarrow v$. Here $u^s(p)$ [$v^s(p)$] is a Dirac spinor describing a particle [antiparticle] with arbitrary spin s. For QED, explain why these equalities are related to U(1) gauge invariance.

(c) Show like in part (b) that $\bar{u}^{s_1}(p)\gamma^\mu\gamma_5 v^{s_2}(q)(p_\mu + q_\mu) = 2m_f\bar{u}^{s_1}(p) \gamma_5 v^{s_2}(q)$ and derive an analogous formula for $u \leftrightarrow v$. Here m_f stands for the fermion mass.

(d) The typical energy scale in the LEP experiment is M_Z. This means that m_e and m_μ can be neglected. Show that this implies that in part (a) the Z propagator can be replaced by

$$\frac{-g_{\mu\nu}}{k^2 - M_Z^2 + i\epsilon}.$$

(e) In the lecture notes (pp. 59, 70) it is explained that quantum corrections modify the propagator. Show for the present case that the Z propagator will be modified to

$$\frac{-Z_Z g_{\mu\nu}}{k^2 - M_Z^2 + iM_Z\Gamma_Z + i\epsilon}$$

($Z_Z =$ Z's wave-function renormalisation; $\Gamma_Z =$ Z's total decay rate).

To a good approximation the k dependence of Γ_Z may be neglected near the resonance. Furthermore, $\mathcal{O}(e^2)$ corrections to Z_Z will be neglected too, i.e., we take $Z_Z = 1$. Why is $\Gamma_Z \neq 0$?

(f) In your calculation below you may (or rather should) use the result from Problem 29(b), namely that in QED the total cross section for $e^+e^- \rightarrow \mu^+\mu^-$ (i.e., via a photon) equals $\frac{\pi\alpha^2}{3E^2}$. Here α is the fine structure constant, the fermion masses are neglected, and the incoming particles are *not* polarised.

Show that the total cross section σ for $e^+e^- \rightarrow \mu^+\mu^-$ via a Z particle equals, for unpolarised electron–positron bundles and in the approximations discussed before,

$$\sigma = \frac{\frac{1}{3}\pi\alpha^2(4E/3)^2}{\left((2E)^2 - M_Z^2\right)^2 + M_Z^2\Gamma_Z^2}.$$

Remark: Since $\Gamma_Z \ll M_Z$ (see below) it is clear from the above formulas that near the resonance photon 'exchange' can be neglected. Also Higgs 'exchange', possible in the standard model, is negligible, as the coupling between the Higgs particle and fermions is proportional to m_f/M_Z ($\ll e$).

(g) Figure 1 on page viii shows the LEP data (in the figure Energy = $2E$). Explain this plot qualitatively from your calculations, and extract Γ_Z.

Note that each fermion–antifermion pair, into which the Z particle can decay will give a positive contribution to Γ_Z. As also neutrinos contribute, one has been able to determine the existence of precisely three (light) neutrino types.

38. **One-loop calculation with scalar fields**
A model with scalar fields φ_0, φ_1 and φ_2 is described by the Lagrangian

$$\mathcal{L} = \tfrac{1}{2}(\partial_\mu\varphi_0\partial^\mu\varphi_0 + \partial_\mu\varphi_1\partial^\mu\varphi_1 + \partial_\mu\varphi_2\partial^\mu\varphi_2 - m_0^2\varphi_0^2 - m_1^2\varphi_1^2 \\ - M^2\varphi_2^2 - \lambda_0\varphi_0\varphi_2^2 - \lambda_1\varphi_1\varphi_2^2),$$

with $M \gg m_1 > 3m_0$.

(a) Even though there is no direct interaction between φ_0 and φ_1, the model gives rise to diagrams with only external φ_0 and φ_1 lines. Clarify this statement by drawing some diagrams contributing to the processes $\varphi_1 \to \varphi_0\varphi_0\varphi_0$ and $\varphi_1\varphi_0 \to \varphi_1\varphi_0$. Use the following notation for the propagators:

$\varphi_0-\ -\ -\ -$ φ_1——— φ_2———

(b) Consider the diagram

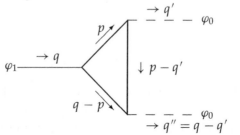

Give the associated S-matrix element without working out the d^4p integration yet.

(c) The S-matrix element contains the following expression:

$$g(q,q') \\ = \int \frac{d^4p}{(2\pi)^4} \frac{i\lambda_1\lambda_0^2}{(p^2 - M^2 + i\varepsilon)((p-q)^2 - M^2 + i\varepsilon)((p-q')^2 - M^2 + i\varepsilon)}.$$

Argue that $g(q,q')$ can be viewed as the effective coupling constant for the leading order contribution to the process $\varphi_1 \to \varphi_0\varphi_0$. Do you expect $g(q,q')$ to be real or complex?

(d) Compute $g(q,q')$ with the techniques introduced in the lecture notes pp. 153–156: write $g(q,q')$ as an integral over a function of the form $I_{n,\alpha,\beta}(\tilde{m})$. To this purpose use a Wick rotation, the Feynman trick, and a shift of the integration variable p. Write the relevant function $I_{n,\alpha,\beta}(\tilde{m})$ in terms of Gamma functions.

(e) Assuming q, q', q'' on-shell, compute q^2, q'^2 and $q \cdot q'$, and observe that these scalars are much smaller than M^2. Use this observation to expand your result from part (d) in terms of $1/M^2$. In this way show that

$$g(q, q') = \frac{g_0}{M^2} + \frac{c_0 m_0^2}{M^4} + \frac{c_1 m_1^2}{M^4} + \mathcal{O}\left(\frac{m^4}{M^6}\right)$$

and calculate g_0, c_0, c_1.

39. **Vacuum polarisation and Pauli–Villars regularisation**

QED, quantum electrodynamics, is the field theory of minimally coupled photons and electrons. Their fields are a U(1) vector field A_μ and a spinor field $\psi(x)$, governed by the Lagrangian

$$\mathcal{L}_{\text{QED}}[A, \psi] = \mathcal{L}_{\text{photon}} + \mathcal{L}_{\text{electron}} + \mathcal{L}_{\text{int}} =$$
$$= -\tfrac{1}{4}F_{\mu\nu}F^{\mu\nu} + \bar{\psi}(i\gamma^\mu \partial_\mu - m)\psi + e A_\mu(\bar{\psi}\gamma^\mu\psi).$$

Choosing the Landau gauge ($\alpha \to \infty$, lecture notes p. 23) the two-point function reads to 0th order

$$\overset{q}{\underset{\mu \nu}{\sim\!\sim\!\sim}} \equiv \Pi_{\mu\nu}^{(0)}(q) = (-i)\left(\frac{-1}{(q^2 + i\varepsilon)^2}[q^2 g_{\mu\nu} - q_\mu q_\nu]\right).$$

[the propagator equals $i\Pi_{\mu\nu}^{(0)}(q)$; compare this to Problem 13]. We are interested in the leading correction to $i\Pi_{\mu\nu}^{(0)}(q)$, the so-called vacuum polarisation:

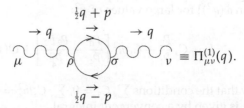

(a) Show that

$$\Pi_{\mu\nu}^{(1)}(q) = \Pi_{\mu\rho}^{(0)}(q)\omega^{\rho\sigma}(q)\Pi_{\sigma\nu}^{(0)}(q),$$

with

$$\omega^{\rho\sigma}(q) = -\frac{e^2}{(2\pi)^4}\int d^4p \, \text{Tr}\left(\gamma^\sigma \frac{m + \gamma \cdot (\tfrac{1}{2}q + p)}{(\tfrac{1}{2}q + p)^2 - m^2 + i\varepsilon}\right.$$
$$\left. \times \gamma^\rho \frac{m + \gamma \cdot (p - \tfrac{1}{2}q)}{(p - \tfrac{1}{2}q)^2 - m^2 + i\varepsilon}\right).$$

Note that this expression is divergent.

(b) Compute the trace in the above expression using your results from Problem 21.

N.B.: Throughout the present exercise do <u>not</u> assume q on-shell.

(c) Show that $q_\rho \omega^{\rho\sigma}(q) = 0$ [strictly speaking this is only valid after regularisation as in part (e)]. From this conclude that $\omega^{\rho\sigma}(q)$ takes the form

$$\omega^{\rho\sigma}(q) = \omega(q^2)(q^2 g^{\rho\sigma} - q^\rho q^\sigma).$$

(d) Determine the scalar function $\omega(q^2)$ by contracting the above expression with $g_{\rho\sigma}$.

(e) The result in part (d) contains a divergent $d^4 p$ integration. This divergence will now be regularised by the method of Pauli–Villars. In this method attention shifts from $\omega(q^2, m) \equiv \omega(q^2)$ to the sum

$$\bar\omega(q^2) \equiv \sum_{s\geq 0} C_s \omega(q^2, m_s),$$

where $C_0 = 1$ and $m_0 = m$. Here the sum should be performed before doing the $d^4 p$ integration.

Show that $\bar\omega(q^2)$ is of the form

$$\bar\omega(q^2) = \int d^4 p \sum_s C_s \frac{P_2 + m_s^2 P_0}{\tilde P_4 + m_s^2 \tilde P_2 + m_s^4 \tilde P_0 + i\varepsilon},$$

where P_n and $\tilde P_n$ are polynomials of degree n in p and of arbitrary degree in q; furthermore they are independent of m_s.

Expand the quotient I_s [appearing within the sum and integration in $\bar\omega(q^2)$] for large values of p^2:

$$I_s = C_s \frac{P_2}{\tilde P_4} + C_s m_s^2 \left[\frac{P_0}{\tilde P_4} - \frac{P_2 \tilde P_2}{\tilde P_4^2} \right] + \mathcal{O}(p^{-6}).$$

Show that the conditions $\sum_s C_s = 0$, $\sum_s C_s m_s^2 = 0$ guarantee that $\bar\omega(q^2)$ is given by a convergent integral.

Remark: Closer inspection shows that the second term in I_s is only of order p^{-6}. The naive conclusion that the condition $\sum_s C_s m_s^2 = 0$ is superfluous is wrong, as the cancellation of the p^{-4} contribution does not take place at the level of $\omega^{\rho\sigma}$. Hence leaving out the second condition makes the derivation in part (c) invalid.

(f) A solution to the conditions is

$$C_0 = 1, \qquad C_1 = 1, \qquad C_2 = -2,$$
$$m_0^2 = m^2, \quad m_1^2 = m^2 + 2\Lambda^2, \quad m_2^2 = m^2 + \Lambda^2,$$

for arbitrary Λ^2. Show that for this choice $\bar{\omega}(q^2)$ gives the same vacuum polarisation as a model consisting of the photon field and the following fields:

	(mass)^2	statistics
ψ_0	m^2	Fermi
ψ_1	$m^2 + 2\Lambda^2$	Fermi
ψ_2	$m^2 + \Lambda^2$	Bose
ψ_3	$m^2 + \Lambda^2$	Bose

with a Lagrangian

$$\mathcal{L} = \mathcal{L}_{\text{photon}} + \sum_{s=0}^{3} \bar{\psi}_s \left[i\gamma^\mu (\partial_\mu - ie A_\mu) + m_s \right] \psi_s.$$

(g) **Remark**: The situation in part (f) describes the regularised theory. To return to the true theory we would like to eliminate the fields $\psi_{1,2,3}$ by pushing their masses to infinity, i.e., by taking the limit $\Lambda^2 \to \infty$. However, careful inspection shows that $\bar{\omega}(q^2)$ diverges as a function of Λ^2:

$$\bar{\omega}(q^2, \Lambda^2) \overset{\Lambda^2 \to \infty}{\sim} \log\left(\frac{\Lambda^2}{m^2}\right).$$

This divergence can be absorbed in a wave-function renormalisation, which will never appear in physical quantities.

40. **Beta decay of the neutron**
Through the weak interactions a neutron (N) can decay in a proton (P), an electron (e) and an antineutrino ($\bar{\nu}^e$). At quark level, this so-called beta decay reads $d \to u + e + \bar{\nu}^e$. The following interaction term in the Lagrangian of the standard model is relevant to this decay:

$$\mathcal{L}_{\text{int}} = \frac{g}{\sqrt{2}} \left(W_\mu^- \bar{\psi}_d^L \gamma^\mu \psi_u^L + W_\mu^- \bar{\psi}_e^L \gamma^\mu \psi_{\nu^e} + h.c. \right)$$

($h.c.$ = Hermitian conjugate, $\psi^L = \frac{1-\gamma_5}{2}\psi$).

(a) Give the lowest order Feynman diagram for the above process (for quarks).

(b) Show that if the external momenta are much smaller than the W boson mass M_W, we can just as well consider the effective interaction

$$\mathcal{L}_{\text{int}}^{\text{eff}} = -\frac{G}{\sqrt{2}} \left(\bar{\psi}_d \gamma^\mu (1 - \gamma_5) \psi_u \bar{\psi}_{\nu^e} \gamma_\mu (1 - \gamma_5) \psi_e \right) + h.c. \quad (23.16)$$

and express the so-called Fermi-constant G in terms of g and M_W.

(c) Prove that $S_{\text{int}}^{\text{eff}} = \int d^4x \mathcal{L}_{\text{int}}^{\text{eff}}$ is <u>not</u> invariant under parity.

(d) Since the proton and neutron are built out of three quarks ($N = ddu$, $P = uud$), one can derive from eq. (23.16) an effective Lagrangian for neutron decay.

$$\tilde{\mathcal{L}}_{\text{int}}^{\text{eff}} = -\frac{G}{\sqrt{2}} \left(\bar{\psi}_N \gamma^\mu (1 - \alpha\gamma_5) \psi_P \, \bar{\psi}_{\nu^e} \gamma_\mu (1 - \gamma_5) \psi_e \right) + h.c.$$

(23.17)

Through QCD-effects α will deviate from 1. In good approximation one has $\alpha = 1.22$.

Give the reduced matrix element $\mathcal{M}$ for the decay of the neutron. Use the following conventions for the momenta (p, k_i) and spins (s, t_i):

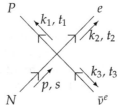

(e) During the beta decay of the neutron, which is assumed at rest, only the momentum of the electron ($\vec{k}_2$) is measured. Using a magnetic field the spin of the neutron is aligned along the positive z axis.

Give the expression for the spinor u_N for this polarisation and prove that

$$\sum_{t_1, t_2, t_3} |\mathcal{M}|^2 = \frac{G^2}{2} \bar{u}_N \gamma^\mu (1 - \alpha\gamma_5)(\slashed{k}_1 + m_P)\gamma^\nu (1 - \alpha\gamma_5)u_N \cdot$$

$$\text{Tr}\left((\slashed{k}_2 + m_e)\gamma_\nu (1 - \gamma_5)\slashed{k}_3\gamma_\mu (1 - \gamma_5)\right).$$
(23.18)

Here m_e, m_P and m_N are the masses of resp. the electron, proton and neutron. We work in the limit $m_N \to \infty$, $m_P \to \infty$, but we keep $\Delta m = m_N - m_P$ fixed.

(f) Show that in this limit

$$\bar{u}_N \gamma^\mu (1 - \alpha\gamma_5)(\slashed{k}_1 + m_P)\gamma^\nu (1 - \alpha\gamma_5)u_N$$
$$= 4m_P^2 \left(c^\mu g^{\mu\nu} - \alpha(\delta_0^\mu \delta_3^\nu + \delta_3^\mu \delta_0^\nu) - i\alpha^2 \varepsilon^{0\mu\nu3} \right),$$

with $c^\mu = 1$ for $\mu = 0$ and $c^\mu = -\alpha^2$ for $\mu = 1, 2, 3$; $\varepsilon^{0123} = -1$, $\varepsilon^{\mu\nu\rho\sigma}$ completely antisymmetric; g is the metric diag$(1, -1, -1, -1)$.
Prove that the 'partial' decay width is given by

$$d\Gamma_\uparrow = f(|\vec{k}_2|) \left(1 - \frac{2\alpha(\alpha - 1)}{1 + 3\alpha^2} \frac{|\vec{k}_2|}{(k_2)^0} \cos\theta \right) d^3k_2,$$
(23.19)

where θ is the angle with the positive z axis, along which the electron is detected.

Hint: Prove first that in the limit $m_N, m_P \to \infty$ conservation of energy implies that

$$\Delta m = |\vec{k}_3| + \sqrt{m_e^2 + |\vec{k}_2|^2}.$$

(g) Explain why the unpolarised partial decay width is given by

$$d\bar{\Gamma} = f(|\vec{k}_2|)d^3k_2 \quad ?$$

Compute from this the lifetime of the neutron in the approximation that $m_e/\Delta m = 0$ (in reality $m_e/\Delta m \approx 0.4$, which leads roughly to a correction with a factor 2). In units where $\hbar = c = 1$, you may use that

$$\Delta m = 2.0 \cdot 10^{21} \mathrm{s}^{-1},$$

$$\frac{m_P}{\Delta m} = 7.3 \cdot 10^2,$$

$$G = 1.0 \cdot 10^{-5} m_P^{-2}.$$

(h) Already in 1957 (breaking of) invariance under parity in the weak interactions was tested. Free neutrons are experimentally hard to handle. This is why a piece of cobalt (Co^{60}) was used, whose nucleus changes under beta decay into nickel (Ni^{60}). Schematically the following result was obtained:

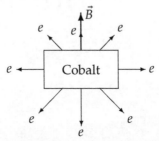

(that is a bigger electron flux in the direction of $-\vec{B}$ than in the direction of $\vec{B}$, where $\vec{B}$ is the applied magnetic field).

Argue why this experiment demonstrated the violation of invariance under parity transformations.

NOTE: Nuclear complications make the precise computations for Co^{60} rather more difficult than those for a free neutron. The result is nevertheless given by eq. (23.19), but with appropriately modified α. For the above question this is not relevant; to conclude that parity invariance (mirror symmetry) is broken, the details of the underlying theory are not required.

Index

Wave functional, 5
  operator action on, 7
Wave packets, 65–66
Wavefunction renormalisation, 59–60
  for fermions, 102
Weak interaction in the standard
        model, Problem 30, 195
Weak interactions
  invariance under parity broken by,
        146
  parity is not a symmetry, 79
Weak mixing angle, 145
Weinberg angle, 145
Weyl representation, 75–77, 79
Wick rotation, 154
Wiener measure, 32
WKB approximation, 38
Would-be Goldstone boson, 135

Y

Yang-Mills equations, 131–132
Yukawa coupling, 150, 153
  constant, 146
Yukawa interaction, 97–98

Z

Z_0 particle creation, vii
Z_0 particle decay, vii
  probability, vii
Z_0 vector boson, vii
Zero occupation number, 6–7
Zero-point energy of vacuum, 7, 8
Zero-point fluctuations, 7, 152
  experimental measurement of, 9
  hyperfine splittings, 9
  Lamb shift in atomic spectra, 9

For Product Safety Concerns and Information please contact our
EU representative GPSR@taylorandfrancis.com Taylor & Francis
Verlag GmbH, Kaufingerstraße 24, 80331 München, Germany